THE SHOTOVER RIVER – 'THE RICHEST RIVER IN THE WORLD'

A History of Gold Mining on the Shotover River

A J De La Mare

GW00566213

Lakes District Museum, 1993

ISBN 0-473-02381-4

Copies available from Craig Printing Co Ltd
67 Tay Street
Invercargill

First Impression 1993-108536
Second Impression 1999-141274

The Lakes District Museum, Arrowtown, gratefully acknowledges the financial assistance given towards the publication costs of this book by the following tourist operators on the Shotover River:

Shotover Jet Ltd
A J Hackett Bungy Ltd
Danes Shotover Rafts Ltd
Mount Cook Lines Ltd

Preface

The history of gold mining on the Shotover River is essentially the history of gold mining in the Wakatipu. For almost 130 years the river has been wrought by every conceivable method of gold recovery, yielding untold riches for some, but disappointment for many. The early miners called the Shotover 'The Richest River in the World', an apt description when fortunes were being won daily. The river still produces a rich return, but today it is from tourism, not gold.

In writing this history, I gathered information from many sources, the principal one being the records of the Lake County Council. Most of these, including rate records and letter books, are held by the Lakes District Museum in Arrowtown. The minute books are held by the Queenstown Lakes District Council, Queenstown. The files of the *Southland Times*, Invercargill, have provided much detailed information as, to a lesser extent, have the files of the *Wakatip Mail*, which are held in Dunedin.

I appreciate the valuable assistance given to me by The Hocken Library, Dunedin, the Invercargill Public Library, the Department of Conservation, Queenstown, the Lakes District Museum, Arrowtown, and Mr G J Griffiths of Dunedin.

I have retained imperial units of measurement in the text, as they are compatible with the events described. Metric measures would appear quite foreign.

The excellent maps are the work of Peter De La Mare, Queenstown. The black and white photographs were provided by the Lakes District Museum, except for those otherwise acknowledged.

In the text I give the height of the Skippers Bridge above the river as approximately 300 feet. Mr Jim Shaw of Maori Point maintains that the Skippers Bridge is now about 220 feet (67 metres) above the river. It can only be assumed that the vast quantity of tailings sluiced into the river over the years has significantly raised the river's bed.

I am grateful to the Lakes District Museum for publishing this history, which is done without profit or remuneration to myself.

A J De La Mare
Queenstown, 1993

The Upper Shotover Management Committee

Late in 1993, as publication date for this book approaches, an important development has taken place that will have a significant effect on the area covered by this history.

In 1992, the principals of A J Hackett Bungy Ltd, recognising that the increasing commercial recreational use of the river was having a detrimental effect on the road and environment, produced a report describing a management strategy for the area.

From this initiative sprang the Upper Shotover Management Committee, as a committee of the Lakes District Council, charged with ensuring that resource management in the area is integrated and with funding and undertaking environmental projects.

The committee is now operational and consists of representatives of Queenstown Lakes District Council, the Department of Conservation, private land owners, miners and tourist operators in the area. Its fundamental goal is the sustainable and integrated management of the Upper Shotover Catchment area, whereby its natural and historical attributes are recognised, conserved, preserved and enhanced in a way in which current use will not compromise use of the area by future generations.

Contents

> A summary of distances and points of interest on the
> road to Skippers can be found on pages 47 – 52.

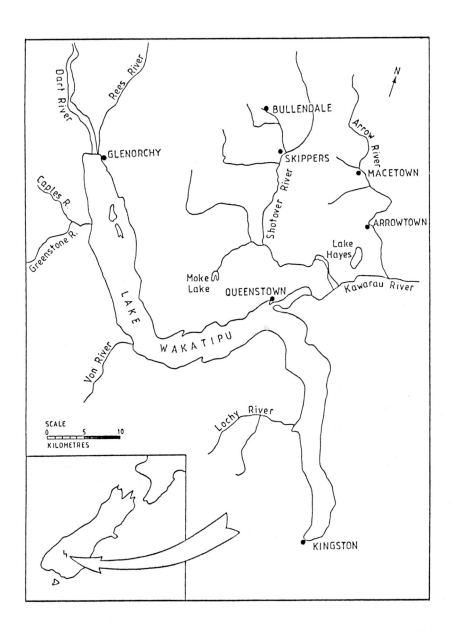

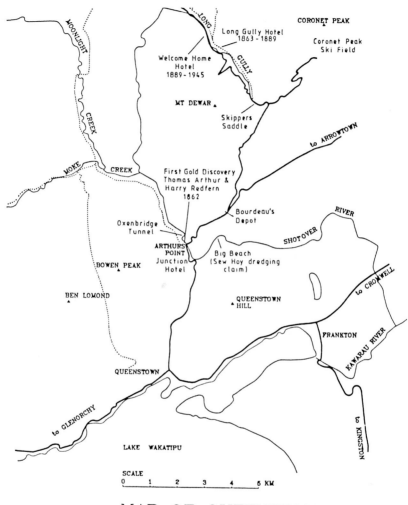

MOONLIGHT

CORONET PEAK

Long Gully Hotel
1863-1889

Coronet Peak
Ski Field

Welcome Home
Hotel
1889-1945

MT DEWAR

CREEK

Skippers
Saddle

to ARROWTOWN

MOKE

CREEK

First Gold Discovery
Thomas Arthur &
Harry Redfern
1862

Bourdeau's
Depot

RIVER

Oxenbridge
Tunnel

SHOTOVER

ARTHURS
POINT
Junction
Hotel

Big Beach
(Sew Hoy dredging
claim)

to CROMWELL

BOWEN PEAK

BEN LOMOND

QUEENSTOWN
HILL

KAWARAU RIVER

FRANKTON

QUEENSTOWN

to GLENORCHY

to KINGSTON

LAKE WAKATIPU

SCALE

0   1   2   3   4   5 KM

MAP OF QUEENSTOWN
AND LOWER SHOTOVER

MAP OF SKIPPERS
AND UPPER SHOTOVER

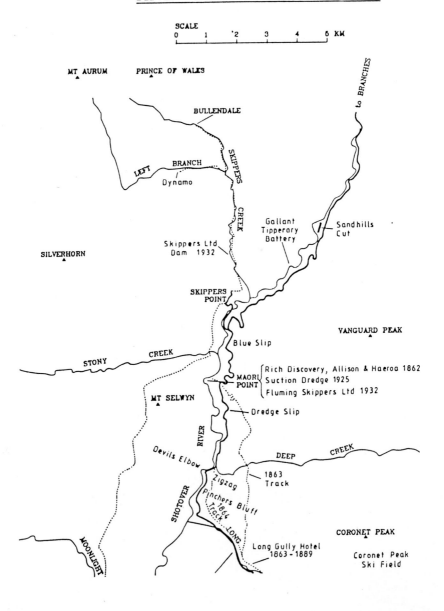

SCALE

0   1   '2   3   4   5   KM

MT AURUM   PRINCE OF WALES

to BRANCHES

BULLENDALE

LEFT   BRANCH

Dynamo

SKIPPERS   CREEK

Gallant
Tipperary   Sandhills
Battery   Cut

Skippers Ltd
Dam 1932

SILVERHORN

SKIPPERS
POINT

VANGUARD PEAK

Blue Slip

STONY   CREEK

MAORI
POINT

Rich Discovery, Allison & Haeroa 1862
Suction Dredge 1925
Fluming Skippers Ltd 1932

MT SELWYN

Dredge Slip

SHOTOVER   RIVER

Devils Elbow

DEEP   CREEK

1863
Track

Zigzag

Pinchers Bluff

1864 Track

LONG

MOONLIGHT

Long Gully Hotel
1863 - 1889

CORONET PEAK

Coronet Peak
Ski Field

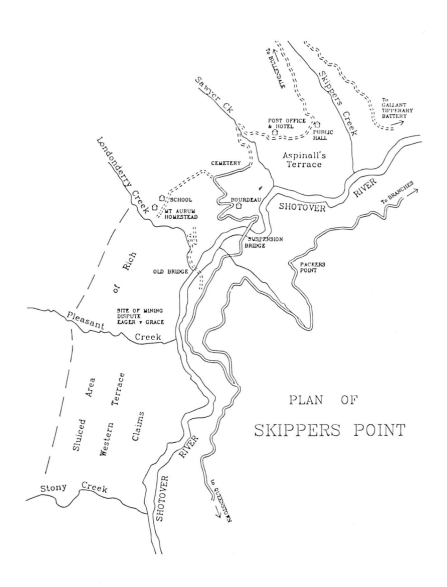

To BULLENDALE

Skippers Creek

To GALLANT TIPPERARY BATTERY

Sawyer Ck

POST OFFICE & HOTEL

PUBLIC HALL

CEMETERY

Aspinall's Terrace

Londonderry Creek

SHOTOVER RIVER

To BRANCHES

SCHOOL

MT AURUM HOMESTEAD

BOURDEAU

SHOTOVER

SUSPENSION BRIDGE

of Rich

OLD BRIDGE

PACKERS POINT

Pleasant

SITE OF MINING DISPUTE EAGER v GRACE

Creek

Sluiced   Area   Western   Terrace   Claims

SHOTOVER RIVER

PLAN OF

SKIPPERS POINT

Stony   Creek

to QUEENSTOWN

1 The Road to Skippers

The rush to the Shotover followed the discovery of gold at Arthurs Point by Thomas Arthur and Harry Redfern in November 1862. Unlike others before them, these two men made no attempt to keep their find secret. As a result, word of their strike soon spread throughout the Wakatipu Basin and beyond, in the same incredibly swift manner as such news always travelled in a gold rush atmosphere. Soon hordes of miners were on the scene. They came knowing that to succeed they had firstly to find a likely claim and to be there before the many others seeking their fortunes.

What they found was a swift, turbulent river issuing from a mountain gorge, the sides of which were precipitous, rocky and reached up to 600 feet in many places. The thought of prospecting such a grim area was daunting, but the lure of gold spurred them on. Climbing the high ground on both sides of the river, the miners pushed on, seeking to make a rich strike. Before long, every beach and the numerous small streams which ran into the Shotover were pegged off, and the once empty mountains were filled with men and frenzied activity.

Working a claim in such a barren and mountainous area was impossible without support to pack in the essential supplies. Initially, men worked in groups, with at least half the partners bringing in supplies on their backs. However, supplying miners' needs was profitable, in many cases more so than mining, and soon store-keepers were packing supplies to establish stores at centres of activity. These were rapidly joined, and soon outnumbered, by hotels supplying liquor and entertainment.

Countless men and horses quickly wore tracks, but as the number of miners increased to several thousands over the entire river system, the need for proper pack tracks became urgent. In June 1863 the *Wakatip Mail* reported that the population of the Maori Point, Skippers Point area was estimated to be 2,000. As this figure was probably arrived at by the editor sitting in his office, its accuracy is doubtful. The population estimate put forward in the same month by W C Wright, a mining surveyor employed by the Provincial Government, is probably the more reliable. His estimate of the population of the Shotover area was 4,116. Wright indicated that he had much difficulty in obtaining the information, but he believed it to be tolerably correct. Given that absolute accuracy is an essential requirement of his profession, it seems reasonable to accept his estimate, which is the only definitive figure of the population at the time of the rush that we now have. It also gives us an indication of the total population of the Wakatipu Basin at this time. If we assume that there were an equal number of miners on the Arrow, and allow for two to three thousands in the towns and other mining areas, we can determine that the total population would have been 10,000 to 12,000.

The quantity of stores needed was huge, and storms in the mountains could easily isolate an area by wiping out temporary tracks. The Provincial Government moved quickly to provide pack tracks to meet the need. Although the entire area was affected by the frenzy and excitement of the gold rush, contractors and labourers were readily available to carry out these tasks. Businessmen of a wide range of skills were prepared to tender for work, even if their experience was limited, and men who had failed to realise their dreams of riches in a gold strike were eager to accept work to avoid destitution.

Engineers surveyed and pegged a track designed for permanence. A contract for the track, following the eastern side of the Shotover and leading from Arthurs Point to Maori Point, was let in July 1863. The track largely followed the route of the existing road to the saddle, then dipped down the eastern side of Long Gully. From the hotel at the bottom of Long Gully the track climbed steeply to Green Gate Saddle, followed Green Gate Creek to Deep Creek, thence to Maori Point along a line much higher than the existing road. The steep climb up to Green Gate Saddle proved to be extremely difficult for horse and man, and in 1864 the track was resurveyed to provide easier grades. Two contracts were let, one to John Braithwaite in June at £1.2.0 per chain, and a second in September to John and Charles Mace at £1.10.0 per chain.

From the bottom of Long Gully the new track went westward of the old over the massive rock feature immediately before Deep Creek. Here, the track descended by a steep zigzag to Deep Creek, and then followed the existing road line to Maori Point. The climb over the rock feature and the steep zigzag to Deep Creek were almost as big a problem as the old track, and it was only overcome at great expense some 20 years later when the road was blasted out of a rock face now known as Pinchers Bluff. The zigzag was a particularly difficult section, with a fall, in parts, of one in three and an almost equivalent grade across the track to the outer edge. It was difficult enough for pack horses to negotiate, but when attempted with heavy loads on trolleys, the track became an almost insurmountable obstacle. Progress was made at a snail's pace and involved real danger for men and horses.

In 1866 the track was carried up the eastern bank of the Shotover to a point opposite Londonderry Creek, where a suspension bridge suitable to carry pack horses was erected over the river. The approaches to this bridge can still be seen clearly today at the junction of Londonderry Creek and the Shotover River.

From here a track led up Skippers Creek to the Bullendale Reefs. It followed the creek bed, except where occasional narrow gorges made it necessary to divert to the hills on the left or right. This narrow track, high above the river, formed the only access to the area for over 20 years. In the initial flush of the gold rush, traffic on the track was immense. Large shanty towns sprang up at Maori Point and Skippers Point and for a brief period were extremely busy. Here, stores providing the miners' essential needs were established. Hotels providing liquor and entertainment also appeared, as well as traders, such as blacksmiths, bakers and butchers. Both towns had a police station, while Maori Point had a Resident Magistrate's Court and a branch of the Bank of New Zealand.

The easily won gold soon petered out, and those seeking quick riches departed, leaving the scene to those content to work a claim with less spectacular but more permanent results. In December 1864 the population in the Maori Point, Skippers Point area was reduced to 410, only 17 of whom were women. About the same time the Provincial Government announced the cancellation of six publicans' licenses at Maori Point, the parties having left the district. The population continued to decrease to the point where, when rating for the Lake County Council commenced in April 1877, the properties on the roll included 76 huts, 35 houses and several stores and hotels with a population of probably two to three hundred people. The lifeline for this population was still the narrow track, with every item laboriously carried in by pack horse. Though the population was small, the demand for better access increased as different methods of gold recovery were introduced. Sluicing required pipes and other heavy equipment, while quartz mining demanded extremely heavy plant for stamper batteries and the like. Packers were ingenious in what they could pack in by horse, but pack horses had to have balanced loads, and there was a limit to the weight they could carry. Bringing in heavy items by specially designed trolleys or sledges provided one answer to the problem, but the incredible difficulties and high cost of maintaining the population and their endeavours by means of a narrow, fragile bridle track remained.

As the years passed, the population of the Shotover Basin became more and more dissatisfied with their means of access. Their concerns were reinforced by the gradual build up in the activities in quartz mining at Bullendale, which presented almost insurmountable problems to getting plant on to the site.

The first positive move came in July 1882 when John Aspinall, riding member for the Shotover, advocated at a meeting of the Lake County Council the great need for a dray road to Skippers. As an illustration of the difficulties and frustrations experienced by those who lived in the area, he said that recently it had taken 70 men to transport a sick person to hospital. Aspinall was well qualified to speak on the problems, as he had lived at Skippers Point for almost 20 years, operating a sluicing claim and bringing up a family. He wasted no time and at the next meeting, some six weeks later, presented a petition that had been drawn up after a large public meeting at Skippers Point. The petition sought improvements to the track at Maori Point and Fergusons Point, and a dray road from Arthurs Point to Butchers Creek some two miles up Skippers Creek. Frederick Evans, manager of the Phoenix Mine at Bullendale, was present and was permitted to speak in support. He said it was impossible to cart heavy pieces of equipment to the Reefs owing to the great risks and enormous expense. He pointed out that £30,000 had been invested at Bullendale to date, all of which was in jeopardy because of access. As the Phoenix Mine contributed significant revenue to the County and contributed much to the economy of the area, his representations were listened to with care. The Council decided a dray road should be formed from Arthurs Point to Bells accommodation house at the bottom of Long Gully and that the local member of the House of Representatives, Thomas Fergus, be asked to obtain a government grant.

With the decision made, there was now urgency to obtain the benefits under the Road and Bridges Construction Act 1881, which had a deadline of the end of the year. The Council moved quickly, engaging a Dunedin civil engineer, James Mollinson, to carry out the necessary planning work. Incredibly, he completed the survey, prepared plans and had estimates for the section Deep Creek to Skippers ready for dispatch five days before the end of the year. Unfortunately, the plans went astray in the post, resulting in the application not being considered under the Roads and Bridges Construction Act. The Government did, however, allocate a grant of £3,400 for the Deep Creek-Skippers section and £3,000 for the Arthurs Point-Deep Creek section. The Government commissioned Mollinson to survey the latter section.

In February 1883, William Rolleston, Minister of Lands, and the Secretary for the Gold Fields, James McKerrow, travelled over the line of the road while in the area. One might have expected that, having viewed the extremely difficult terrain the road was to traverse and the small number of people it was to serve, they would have made a move to withdraw the grants. If they were so inclined, they made no such move. The work was finally split into four sections:

Section No. 1   Arthurs Point to the Saddle
Section No. 2   The Saddle to Deep Creek
Section No. 3   Deep Creek to Maori Point
Section No. 4   Maori Point to Fishers.

Fishers was named after Edward Fisher, who operated a store and probably a hotel on the east bank of the Shotover opposite Londonderry Creek.

Tenders were sought for sections 1, 3 and 4, but only one, that from Walsh Davis & Company of £1,923 for section 1, was accepted. In August 1883, with government approval, the work on section 1 commenced and was completed thirteen months later. A tender of £1,660 from Courtenay & Company for section 3 was accepted in September 1883, this work being carried on contemporaneously with section 1. On completion of sections 1 and 3, a tender of £1,749 was accepted from Walsh Davis & Company for the No. 4 section that finished opposite Skippers Point. This section was completed in July 1885.

After three years' work, some nine and a half miles of the planned road had been completed, leaving a section in its middle of almost six miles to be done. The residents of the Skippers area therefore remained reliant on pack horses, and as the most difficult piece of the road had been left to last, the shifting of heavy equipment continued to be difficult. The Council decided that because the zigzag was the main problem area, it should be tackled next. Although it would have been logical to start section 2 at the Saddle and to work towards Deep Creek, the Council considered it essential to eliminate this nightmarish section. This work required a separate contract for a two-mile section of road where 10 chains of solid rock would have to be blasted through. Government approved the suggestion, and the Council accepted a tender of £3,800 from Davis & Company.

The section, now known as Pinchers Bluff and the Devils Elbow, called for a platform for the road to be blasted out of a sheer face which rose some 600 feet out of the river. The task was a daunting one, involving hand-drilling and explosives. Men were lowered on ropes to be suspended like flies on a wall as they carried out endless hand-drilling into the hard rock face. The work was hazardous and difficult, as two brothers by the name of McConochie found. They were suspended over a rock face with ropes around their waists, when, suddenly, the whole rock mass began to move. To hesitate was fatal, and they literally ran up the moving face to safety. They were not a second too soon, for the whole face slid away and down into the river far below. Tourists who today stop on this part of the road to peer cautiously over the edge to the river far below never fail to admire the tenacity and courage of those who laboured here over 100 years ago.

The road around the zigzag was finally completed in October 1886. However, the section from the Saddle to meet the new road at the zigzag remained as a pack track. A shortage of government funds saw this unsatisfactory state of affairs continue for another two years. In November 1888 a contract for section 2 for £3,733.6.7. was let to John Maher and Son of Invercargill. The road to be formed was far from the river and consisted mainly of a long descent of almost three miles down Long Gully, with a drop of approximately 1,300 feet. Today, anyone who stands at the top of the road, watching it snake away to finally disappear at the bottom of the valley far below, is in for a spectacular sight. John Maher employed some 40 men, but he found constructing the road a tougher job than he had expected. Towards the end of his contract, he attempted to ease his financial difficulties by charging a toll to use his partly formed road. He was rebuked by the County for doing so and told to desist. Maher finally finished his contract in May 1890, and returned to Invercargill. Shortly after he was declared bankrupt. He blamed his financial problems on his contract on the Skippers Road.

With Maher's section finished, the road from Arthurs Point to Skippers Point was now completed. Seven years on, 15 miles of narrow road had at last been constructed, the dream of all those who, over the years, had put up with the inconvenience, difficulties and dangers of a pack track. Forcing a road through this precipitous rugged country with modern equipment would not be easy, but with the primitive gear available at the time it was a wonderful achievement. Most of the work was done by pick, shovel and wheelbarrow, with explosives skilfully used in rock areas.

Horse-drawn drays were the only means of transport, and gravel supplies were often long distances from the work face. On many steep faces a road platform could only be achieved by building a rock wall from the nearest foundation area below the road. Dry rock walls, some quite extensive, are still evident today. The workers laboured under extremely difficult conditions. Living in tents in the extreme cold and snow of the long winters was hard enough, but working in the open during these times was a real test of endurance. Work was further complicated by the impossibility of excavating ground frozen solid by severe frosts.

The road remains today as a memorial to the hard work, tenacity and skill of the roadmakers of the past. In 1890, however, it ended at the bridge at Londonderry Creek. With quartz mining concentrated on the west bank of the river, the problems of transporting the heavy plant required in these operations commenced again at this point. The road had no sooner been completed than the clamour for a new bridge began. Quartz mining at Bullendale was reaching a peak, and Fred Evans, the mine manager, was a powerful influence with the County. Planning for a new bridge soon got underway.

To get an acceptable grade it was found that a new bridge on the existing site would mean raising it by 70 feet and lengthening it by 180 feet. The site finally chosen placed the new bridge in a spectacular position. Three hundred and sixteen feet long and set approximately 300 feet above the river, it would span a gorge with sheer rock faces on both sides. The approaches on the western bank would be fairly steep, falling over 200 feet in a short distance to the bridge. Building a bridge in such a position would be costly, and the work was no job for the faint-hearted.

Towards the end of 1893 the Mines Department gave approval to the project and indicated that a subsidy of £1,500 would be forthcoming. As it was obvious that the cost would be much greater than this amount, further consideration was given to the plans. Changes were made reducing the width to eight feet, and the Government finally approved a grant of £2,744. As the lowest tender was £3,244, received from A J Fraser & Company of Naseby, the shortfall of £500, plus the cost of the approaches (£781), had to be found by the County. The Queenstown Borough Council and the Arrowtown Borough Council each contributed £25, the balance being raised by a special loan. The loan was for £2,000, for a term of 26 years. It was secured by a rate over the entire county.

Work on the bridge began in March 1898 and took two years to complete. The bridge is suspended by fourteen 4.5 inch wire ropes, each with a breaking strength of 75 tons. The bridge's official opening on 29 March 1901 by Mr James McGowan, the Minister of Mines, was a tremendous occasion for the residents of the area. The ceremony, which took place at 1.00 p.m., was followed by a banquet in Mrs Johnston's Otago Hotel at 4.30 p.m. and then a ball that night in the old library building. Celebrations on this scale in a small settlement like Skippers placed a strain on the limited facilities. The extremely limited dining room and kitchen facilities would have posed very difficult catering problems for Mrs Johnston, and the library building must have been fairly bursting at the seams to accommodate the even larger attendance at the ball. In addition, most of those present would have required overnight accommodation, despite the fact that, in accordance with the custom of the day, the ball very likely carried on to daylight.

It was a day of triumph for Skippers Point. For almost 40 years access had been difficult, inconvenient and at times dangerous. Now it was at last possible to travel in a horse-driven vehicle from Queenstown right to the Point with, for the day, acceptable speed and comfort. But it was too late. The high cost of building roads and bridges in the difficult country and the Government's never-ending financial problems in the depressed decade of the 1880s meant nothing happened when it was needed. Projects having got into the bureaucratic machine moved ahead slowly and inexorably to completion long after they were wanted. The members of the Lake County Council were driven over the years by the great need for suitable access to the quartz mining at Bullendale, but even when the road and bridge provided this to Skippers Point, there still remained five miles up Skippers Creek negotiable only by pack horse. Further activity in the area was very soon to cease for ever, and all the heavy equipment was in position, having been transported at great effort and expense long since.

Since that red letter day in 1901, the road and bridge have continued to provide access for gold mining, pastoral farming and tourist projects. Maintaining the road over the years has been a never-ending problem for the Lake County Council and its successor, the Queenstown Lakes District Council. Heavy rains in steep, unstable country bring slips and blockages, some requiring major repair work. One section of the road near Skippers Bridge crosses an area called the Blue Slip. The terrain at this point is wet and consists of fragmented rock. It is continually on the move and has been an ongoing problem.

The road is steep, winding and narrow, and for most of its length the country rises abruptly above the road on the inside with an equivalent fall on the outside. Driving on the road, even in the days of horse-drawn vehicles, required extreme care. The hazards were even greater in the snow and ice of winter. The Lake County Council, conscious of the dangers, took several steps over time to promote safety. In 1896 cycles were prohibited on the road. In 1906 the use of motor vehicles was prohibited on a number of country roads, including the Skippers Road. With the passage of time and with more motor vehicles being brought into use, the restrictions were removed, except for the Skippers Road. In 1918, in response to public demand, motor vehicles were permitted on the road between the hours of 7 p.m. and 8 a.m. Further changes were made, until the restrictions were entirely removed in 1926.

For many years there has been a sign at the Saddle before the road starts its journey down Long Gully indicating the road is for 'Experienced Drivers Only'. It no doubt deters many, but some who should be deterred gaily launch themselves on a journey which they later regret starting. Despite the difficult nature of the road, it is busy with tourist buses, rafting vehicles drawing trailers, other commercial vehicles, and many private motorists. The latter, experienced or otherwise, who pass the cautionary sign either enjoy the unique scenery of the Shotover River or return promising themselves never again to venture in their own cars on such a hair-raising ride.

The first Skippers Bridge at Londonderry Creek, 1866.

Skippers Point, showing sluicing creeping up to Mt Aurum homestead and Skippers School on the left. The building on the right is a private dwelling.

Aspinall's claim at the junction of the Shotover River and Skippers Creek. Johnston's Otago Hotel is on the left and the public hall is on the right. Hocken Library

The Otago Hotel at Skippers Point. The license lapsed in 1919 and the building was later dismantled.

Workmen's houses perched high on a ridge above the machine house at Bullendale. The power line from the hydro-electricity scheme is visible on the left.

The Bullendale machine house, containing stamping batteries, the gold recovery plant and workshops. A rock breaker is to the right of the smoke stack.

2 Mining at Bullendale

When the van of the miners reached Skippers Creek late in 1862, they found rich deposits of gold which could be easily won with pan or cradle. Soon the entire creek and surrounding country were pegged off. In mid-1863, Alex Olsen, a Norwegian, discovered quartz that showed great promise in the headwaters of the Creek. One sample returned one ounce of gold from a bucket of stone. His discovery led to the realisation that there were other sources of gold in the area apart from alluvial, and soon many claims were pegged off. In March 1866, W C Wright, Provincial Government Mining Surveyor, reported 18 claims pegged off in the area. Most of these were speculative, with little work apart from prospecting having been carried out.

Quartz mining was, however, a very different proposition to working rivers and streams. It involved underground mining, heavy plant to crush the quartz and, above all, capital to finance the undertaking. The greatest difficulty was the fact that the only access to the claims was an indifferent pack track over 20 miles long. The difficulties and problems of mining in these circumstances soon sorted out the faint-hearted. Only two companies managed to overcome the difficulties and begin mining in earnest. They were the Otago, which included Olsen, the original discoverer, and his partners Andrew Southberg and Alexander Murdoch, and the Scandinavian, owned by Samuel Jones and party. Their stamping batteries were small and primitive. The stamper boxes and every other part of the plant, where practical, were cast in pieces and put together on arrival. The stamps had wooden shanks, with the shoes made by fixing metal to the bases, while the cams to raise the stampers engaged on wooden pegs.

Crushing quartz in a stamper battery requires heavy gear, and wear and tear is a serious problem. The early plant at Bullendale must have been inefficient and probably broken down for long periods. Water from the same source supplied the power for both batteries, and lack of water at times brought crushing to a halt.

After operating for 18 months as a syndicate, a company called the Great Scandinavian Quartz Mining Company Ltd was registered in May

The Shotover jet in the Shotover River canyons.
Shotover Jet Ltd

The Danes raft shooting the rapids on the Shotover River near Oxenbridge Tunnel.
Danes Shotover Rafts Ltd

The ultimate thrill—bungy jumping from Skippers Bridge with A J Hackett Bungy.
A J Hackett Bungy, Queenstown

A Mount Cook Line tourist coach on Skippers Bridge.
Mount Cook Line Ltd

1866. It bought out the miners who were the owners of the Scandinavian claim. With the company came George Francis Bullen, who was to be associated with quartz mining in the area for over 30 years, and after whom the settlement was named. Bullen, a merchant, and his two brothers had operated in the gold mining areas. About this time, the Bullen brothers were disposing of their business and concentrating on pastoral pursuits in the Kaikoura area.

In January 1867 the Otago claim was taken over by the Otago Quartz Mining Company Ltd. The shareholders included the original owners— Olsen, Murdoch and Southberg. Other companies were operating in the area at this time, but they gradually folded until only the Great Scandinavian and the Otago remained. By 1885 only the former was still operating, having absorbed all the other claims.

The arrival of George Bullen brought an increase of activity through the injection of fresh capital. The name Phoenix began to replace Scandinavian as the name of the mine, and Frederick Evans arrived to become mine manager. Evans was engaged as a result of an advertisement in the Melbourne newspapers seeking an experienced mine manager. Born in Cornwall, Evans was to manage the company for 30 years, and as the company was the largest commercial enterprise in the Wakatipu, he exercised considerable influence.

However, the appointment of an experienced mine manager did not bring success. A 30-head stamper, purchased in Melbourne and erected at the mine after enormous difficulties in getting it there, could not be used to its full capacity because there was insufficient water available to power the plant. There was also a shortage of payable stone, although from time to time, patches showing good returns were struck, raising hopes of a return to profitable working. The company prospected to find payable stone and made other changes to improve profitability. It installed new gold-saving machinery and introduced different methods of working the claim, including letting the mine on tribute. But these efforts were to no avail, and for 17 years one poor year followed another until, in 1884, the prospecting paid off when, after sinking a shaft, some good stone was found some 150 feet lower than any previous workings.

This was good news, but there was a twist in it. Working the new area required powerful winding gear for the new shaft and high-capacity pumps to deal with the large body of water encountered at the lower level. It was also evident that water power was unreliable and that some form of steam power was needed to maintain production. The high cost of the additional plant prompted Evans to suggest to Bullen that he float the mine on the London market, the usual source of large speculative capital. Bullen, however, decided to proceed alone, his decision no doubt taken in the hope that the new area of claim would be sufficiently profitable to cover the cost. His gamble paid off. Returns for the 1885 through '87 period were good, helped by the discovery in March 1886 of a promising reef during the cutting of a race to bring in water from the Old Man Creek. The following figures show the good returns enjoyed by the Phoenix Mine after its long time in the doldrums:

26 June 1885	1,003 ozs gold	(600 tons stone)
13 July 1885	330 ozs gold	(200 tons stone)
23 September 1885	700 ozs gold	
30 September 1885	550 ozs gold	
26 November 1885	350 ozs gold	(250 tons stone)
6 January 1886	291 ozs gold	
13 February 1886	1,400 ozs gold	(730 tons stone)
9 July 1886	550 ozs gold	(400 tons stone)
8 November 1886	430 ozs gold	(one month's crushing)
30 December 1886	360 ozs gold	(2 weeks' crushing)
5 April 1887	340 ozs gold	
19 April 1887	200 ozs gold	(10 days' crushing)

These returns were extremely good, some remarkably so, especially that of 13 February 1886.

The good returns enabled Bullen to carry out much new development. A description of machinery installed or in the course of erection at this time illustrates the extent and complexity of plant needed to keep the mine operational:

The crushing machine is one of six batteries, each of five revolving stampers weighing 8 cwt, the batteries having a grating with 149 holes to the square inch. A peculiarity is that no quicksilver whatever is used in saving the gold, except in the amalgam barrel. This allows crushing to be carried on during periods of frost when under ordinary conditions silver would become inactive. In fact here, as elsewhere, the silver process is gradually being discarded. Another peculiarity is that the machine drives the camshaft by a vulcanised belt which is substituted for the usual connecting rod to ensure smooth working and prevent jarring of the machinery, as a fly wheel at Skippers is an undreamed of luxury. The motive power is supplied by a turbine wheel when water is available. At other times a steam engine of 14 horse power drives the stampers.

Self-feeding boxes are provided at all the batteries, and a stone crusher to
reduce all lumps to a given size is now in the course of erection. So far
these self-feeding boxes have given every satisfaction. The whole of the
machinery is covered in a house measuring 45 feet by 85 feet and the
turbine wheel is sunk to a depth of twenty feet below the floor of the
engine house in order to obtain additional pressure. In the machine
house there is also a complete turning lathe adapted for wood, iron and
steel turnings, complete engineer's and carpenter's shop provided with
circular saws, boring machinery, etc which can be driven by either steam,
water or electricity. The whole of the house and all its subdivisions are
lighted by electric lights, the electricity being generated by a pelton
wheel of 2.5 horse-power with a head of 170 feet of pressure. Sufficient
electricity is generated to light the machine house, the library and the
manager's residence with, in all, 20 lights of 16 candle power each. Two
compressed air engines are in the course of erection at the mouth of the
low level adit. where a shaft has been commenced and is now down 82
feet. These engines are of 20 horse-power each and are intended for
haulage, pumping and driving rock drills. The air to work them is
compressed by steam. In addition to this machinery there are two
powerful British dynamos erected in what is known as the Left Hand
Branch Creek, each of which is calculated to generate electricity equal to
96 horse-power. The electricity from this source drives the stamps.

The company introduced electricity to power the main machinery
so as to overcome the short-comings of other sources of energy. Water was
unreliable and steam had its drawbacks—mostly related to the adequacy
of fuel supplies. Electricity answered all the problems.

Late in 1884, Walter Prince, an employee of the pioneer electrical
firm of R E Fletcher of Dunedin, inspected the Bullendale area to
determine the feasibility of using hydro-electricity. In February 1885, a
contract of £2,000 was signed for the erection of generating plant in the
left-hand branch of Skippers Creek, the power to be conveyed to the
machine house by overhead power lines, and the mine to bring in water
by means of a race to drive the pelton wheels which, in turn, would drive
the generators.

The mine development produced other activity. Harry A Evans
opened a combined grocer, baker, butchery business, and James Johnston
built a billiard room, a facility without which any small town was incomplete.
George Bullen also built, at a cost of £150, 'Bullen Hall', thereby providing
the settlement with a library and meeting place. The building, designed
to hold 150 people, was comfortably furnished and had the rare advantage
of being lit with six electric lights.

With some 80 to 90 men now employed in the area, more of the boxes that were the Bullendale equivalent of houses were erected on the high ground behind the stamper house. The period would have been a buoyant one for packers, for although portions of the dray road were completed, the old pack track still served the remaining sections. In March 1885, Fred Evans appealed to the Lake County Council to improve the access, so that heavy plant could be transported to the mine. The items included a three-ton stone-breaker, two dynamos weighing six tons in total, two pelton wheels (each six feet in diameter), 10 tons of castings and 700 feet of large iron pipe. The difficulties of bringing in these extremely heavy items via the dray road were compounded by the fact that it existed only to the Saddle and from Deep Creek to Skippers Bridge. The dreaded zigzag leading down to Deep Creek still existed, as did the bottleneck of a pack bridge at Skippers Point. A stone-breaker could no doubt be sledged through the river by brute force, but a dynamo with parts sensitive to moisture could not. The only way to overcome this obstacle was to completely dismantle the dynamo, taking the delicate parts over the bridge and the balance through the river. But even after arriving at Skippers Point, the five-mile, often difficult, track up the creek to the mine still had to be traversed.

In June, Evans again made representations to the Council, this time in person. He complained of the standard of work on the road, accused the County Inspector of not doing his job and said that his company had spent £50 on improving the track up the creek. The Council repudiated the accusation directed at the inspector, but agreed to subsidise the company pound for pound up to £50 for work done on the track.

The work on installing the hydro scheme, due for completion in 1885, was ready for trial in February 1886. The Phoenix Company brought in water by race from the left-hand branch of Skippers Creek to the top of the bluff, at the base of which stood a building housing the power plant. There was a vertical drop of 185 feet to two pelton wheels, which, in turn, drove the generators by driving belts. The pelton wheels were built by A & G Price of Thames, and the arc dynamos were manufactured by The Brush Company of Australia. Heavy copper wire, mounted on twin poles over a high ridge to the battery house some one and a half miles away, transmitted the power.

The equipment was brought into operational use in March 1886, and the event marked the first use of electricity for industrial purposes in New Zealand. However, the generators supplied only enough power to drive 20 stampers, leaving 10 idle or powered by steam. Nevertheless,

Evans and Bullen were reasonably happy with the result and approved some modifications which, they hoped, would increase the power delivered.

That the scheme worked at all was remarkable given the infant state of electrical generation in New Zealand, and the use that Bullen and Evans put it to at Bullendale broke new ground. These considerations, along with the remoteness and difficulty of access, made the project's successful completion a signal success, but it did not appear to help the company which had planned and carried out the installation. Within two months of completing the job, the firm was bankrupt.

Just before the project neared completion, Walter Prince, the installing electrician, was thrown from his horse in Long Gully. He received serious head injuries, and his life was in danger. His injuries did not prove fatal, but in the words of the day, 'his reason was despaired for'. In time he was taken back to Dunedin, where he made a full recovery. Prince must have been accident-prone. In December 1889, when travelling from Christchurch to Dunedin by train, he fell off a carriage platform while reading a newspaper. He was found unconscious but once again recovered.

The last years of the decade were as bad as those before 1884. In 1888, 2,500 tons of stone were crushed with very poor returns, while in 1889 and 1890 the mine was virtually at a standstill, the only activity being prospecting. At the end of that year the Phoenix Extended Company, which had operated a claim north-west of the Phoenix ground for four years, went into liquidation. The Phoenix Company bought the claim and, in a drive from the shaft in a northerly direction, found good stone. Once again the Phoenix mine was operating successfully, as returns for the last parts of 1890 and 1891 show:

8 October 1890	312 ozs	(600 tons)
6 December 1890	480 ozs	(300 tons of stone crushed in 3 weeks)
28 April 1891	240 ozs	
16 November 1891	690 ozs	(1 month's crushing)
14 December 1891	1000 ozs	(2 months' crushing)

The Mines Department annual report for 1892 records that in 1891, 4,932 tons of stone were crushed for a return of 3,438 ounces of gold, valued at £13,700. Expenses in the same period were £13,500. Although returns were good, profitability was not.

In April 1891, Fred Evans left to visit the United Kingdom, with a view to floating a company to take over the enterprise. Before his departure he was the guest of honour at a banquet put on by his staff, who presented him with an illuminated address and a purse of 50 sovereigns. He said at the dinner the mine was going well, the last cake sent down being 240 ounces. The reason for the move is not known, but probably Bullen, having been involved at Bullendale for just on 25 years without receiving any significant dividends, wanted to try to capitalise on his vast investment in the mine. With the mine showing good returns, it was an opportune time to sell the company.

In 1892, 5,457 tons of stone were crushed to yield 1,920 ounces of gold, and with about 60 men employed, the year would have shown a substantial loss. The stone from the Phoenix Extended shaft that showed such good returns in 1890/91 had petered out. It appears that this news affected Fred Evans' efforts to get financial backing in London, but he did manage to sell the mine to an English company, Achilles Goldfields Ltd, registered in March 1893 with a capital of £100,000.

Bullen was allocated 80,000 paid up shares, while his holding was secured by a mortgage over the claim and plant. Bullen's position was little different to that before the formation of the company except that he now owned shares which he could sell if a buyer was available. Originally, only 2,960 shares, each 2/6d, were taken up by investors, which provided a working capital of a mere £370. By June, only 5,582 shares had been taken up, which showed that the investment was not viewed with enthusiasm. On 30 March 1895, probably as a result of financial or other forms of inducement, the *Illustrated London News* featured a full page on the company, complete with four photographs. The editorial stated that the company owned the celebrated Phoenix Mine, known to be one of the richest in New Zealand. Already some £50,000 had been spent on development. Investors were invited to direct their enquiries to the New Zealand Colonial Office, Victoria Street.

The new company took possession on 16 September 1893, with promises of big developments to come, including the advice that 1,500 tons of equipment was on the way. The promises were not fulfilled for reasons that were not hard to find. Between 16 September 1893 and 24 June 1895, the company crushed 6,854 tons of stone for a return of 2,806 ounces of gold. From November 1895 to June 1896, the returns were much better, with 3,169 ounces of gold from 3,832 tons of stone crushed. This return probably stemmed from striking a rich section of the reef. In December 1895, quartz estimated to yield five ounces of gold per ton of stone was on view in the Bank of New Zealand, Queenstown. A return of over one ounce per ton was considered good, so this stone was exceptionally rich. Fred Evans said he could pick five tons equally as rich from stone mined ready for crushing.   .

In 1896 and the early part of 1897 some reasonable returns were received:

May 1896	560 ozs of gold	(600 tons stone)
July 1896	762 ozs of gold	(750 tons stone)
January 1897	713 ozs of gold	(6 weeks' crushing)
March 1897	578 ozs of gold	(461 tons stone; 13 days' crushing)
May 1897	600 ozs of gold	(2 weeks' crushing)
July 1897	483 ozs of gold	(484 tons stone)

But returns fell off during the second half of 1897, until, by 1898, the company was in debt, with even wages unpaid. The workers, angry at not being paid, took out liens against the company, and at one point the likelihood of the liens being exercised appeared a real threat. To avoid this action and to prevent the mine being sold cheaply, the company required extra capital immediately. At a meeting of shareholders held in London, it was agreed to form a new company to take over the assets. The capital of the new company, which was named the Achilles Gold Mines Ltd, was 92,000 shares, each 2/6d, treated as paid up 1/6d per share. A call of 6d per share was immediately made to meet the company's obligations.

In the same year, Fred Evans retired to his farm at Frankton after having spent over 30 years as manager at Bullendale. His advancing years and the demands of a job in a difficult area were probably the main reasons for his retirement. There was possibly another factor. In March 1897 his house at Bullendale caught fire and it, together with the company office and storerooms, were completely destroyed. Surveying and scientific instruments, mine records, plans and books were all lost. It was a major disaster and probably helped Evans decide it was time to give up after over 30 years as manager at Bullendale. His place was taken by two English engineers, N C Morcom and James Cherry. Morcom did not remain in charge for long, as he suffered a stroke. His assistant, Cherry, took over, but he too departed, to be replaced by F J Donnal. Donnal remained until the mine finally closed in May 1901.

By this time, about 40 men depended on the mine for their livelihoods, and moves were made to take it over on tribute. Apparently, this was not acceptable to the company. Patrick Cotter, whose father owned the general store and in this capacity acted as unofficial banker to many of the miners, offered on their behalf to buy the mine for £2,000. It was generally considered that there was gold to be won in the area. Wesley Turton, a Queenstown solicitor acting for the English owners, indicated that £5,000 was the minimum that would be accepted. As this sum was beyond the means of the miners, the closure went ahead, with everybody having to accept the loss of their jobs.

Finally, in December 1903, Turton paid the Lake County Council arrears of rates of £66.4.3 and rent of £114.15.0 and advised the mine had been sold to R Lee and party. A company was formed by the promoters, one of the shareholders being Joseph Ward—then a Minister of the Crown. However, mining activity for the newly formed Mt Aurum Quartz Mining Company Ltd was short-lived. The mine closed for the winter of 1905 and never re-opened.

The Bullendale mine operated for just on four decades. Many people lived and worked in this remote area for years, some for almost a full working lifetime. It would be fair to say that neither promoters nor workers gained financially from their association with the mine. Indeed, some lost heavily. There was ground in the Bullendale area that held rich quartz, but the broken nature of the country meant that there was no continuity to the reef. The efforts involved in finding it again, or an equivalent productive area, could be a long and costly business. Furthermore, the quality of the quartz in the reefs varied considerably. An area showing excellent returns could suddenly change to quartz not worth the cost of mining and crushing. On the credit side, gold production during the mine's lifetime probably exceeded 35,000 ounces. But the debit side—the expenses associated with extracting the ore—far outweighed the returns.

The road down Long Gully, showing the 1863 pack track in the background.
Bob O'Connor

The view from Maori Point saddle looking towards Skippers Point and showing the western terrace sluicing claims.
Jim Shaw

3 Life at Bullendale

The first miners had penetrated to the headwaters of the Skip–pers Creek by November 1862, and for a little over 40 years there was a settlement in the area called Bullendale. The number living in the area fluctuated depending on the fortunes of the mine, and it is virtually impossible to accurately assess the population at any particular time. One of the reasons for this is that Bullendale residents were usually included as part of Skippers Point in the statistics. Even the early rates records of the Lake County Council do not show Bullendale as an address, only Skippers plus a few shown as The Reefs.

When people arrived in the area to work, their first need was some sort of house to live in, a tent being entirely inadequate in the winter months. There was plenty of beech forest in the vicinity, but until a sawmill was established later in the life of the settlement, pit-sawing the minimum of timber and using galvanised corrugated iron for sheathing was the most practical solution. Galvanised iron sheeting became almost the universal building material on the gold fields. It was easy to use, strong and could be brought in by pack horse. However, it had one tremendous disadvantage. It magnified the heat of the summer and the cold of winter. As the mining area was overshadowed by high ground, the miners erected their huts high on a spur west of the stamper house. In this position they got the most sunshine possible and were a good distance from the noise and drift of dust from the stampers.

The buildings were simple, even primitive, of one or two rooms. An indifferent water supply, no hot water and an outside privy were the norm. Heating and cooking came from an open fire or a black iron range, and an outside copper would be considered a luxury laundry. Many probably had earth floors to begin with. Getting to work, the shop, hotel or school meant going down steep narrow tracks, difficult at night and dangerous when iced up.

In these 'boxes', people lived, some managing to bring up families. Some men spent almost their entire working life at Bullendale. Fred Evans, the manager, was one such. Another who lived and worked at Bullendale for at least 30 years was James Edwards who came from Evans' home town—St Agnes, in Cornwall. Edwards occupied the position of general foreman for many years. As families tended to be large, the customary two rooms soon became inadequate and a lean-to would be added as a simple and quick solution to a burgeoning family. Simple comforts would come as time passed, particularly when a woman was present. For the first 20 years, the number of houses and huts probably numbered no more than 20, but from the mid-1880s, when activity increased, the number probably reached at least 50. A number of men preferred to leave their families in the comparative comfort of nearby towns, and single men mostly used the accommodation provided by the company. At peak times, the total population probably approached 200.

Bullendale sat 2,175 feet above sea level, in a very sheltered area. The summers were delightfully warm, often hot, and the winters extremely harsh. Snow fell frequently, sometimes several feet at a time and, with the continuing severe frosts, never thawed throughout the entire winter. Everybody, including the children, needed spikes in their boots to negotiate the expanses of ice that formed everywhere. Gathering sufficient firewood to last the winter was essential. Wood had to be cut and stored under cover, otherwise it became a block of ice and useless for firing. The winters were a long struggle to keep warm, and hygiene suffered, as for weeks on end drying clothes outside was virtually impossible. Women bringing up families in this climate and under primitive conditions toiled unceasingly. Cooking, washing clothes, sewing, knitting, soap making and the daily chore of milking goats were all part of a busy routine. The plentiful supply of small fruits in the area made preserving in season an additional task.

For supplies there was the combined grocer, baker, butcher, draper, hardware shop. It also served as a post office. This building was the centre of the settlement. Saturday night was mail night, and with it arrived the *Otago Witness*, the only means of knowing what was happening further afield in New Zealand and overseas. All the goods for the shop were packed in from Skippers Point, and meat came from a slaughter house situated on the creek below the settlement. Butchering in the extremes of climate had its problems. In the hot summers the only means of keeping meat was by using a tunnel driven into a bank behind the shop. In the winter, a carcase would have to be thawed in a warm room before it could be cut up.

The combined shop was opened in 1885 by Harry A Evans, who was followed by H Graham in January 1888. Thomas James Cotter and his brother, Richard Joseph Cotter, bought the business in May 1894, remaining till it was closed down in September 1902. The owners, as listed, also acted as postmaster. There was a post office for a short period in 1878-80, the postmaster being Fred Evans. The office was named The Reefs, but this was changed to 'Bullendale' in 1892. A telephone service was established in 1896. The Government had maintained a service to Skippers Point from 1883, and the extension to Bullendale was subject to a guarantee of £32 per annum being met. The guarantee was shared equally by the Lake County Council and the mine management. That the connection received good use is shown by a refund in July 1898 of one fifth of the subsidy, namely £6.80. The revenue had exceeded the minimum laid down.

Another important part of the settlement was the Phoenix Hotel. Built of the inevitable galvanised iron, it was situated on the banks of the creek just below the stamper battery house. It does not appear to have been very well sited. There must have been anxious times when floods either entered or threatened the building, and the nearby stamper house would surely have made it subject to excessive noise and drift of quartz dust. Nevertheless, it was an important social centre. Apart from billiards, there was little entertainment, and drinking in the hotel was the usual relaxation for most of the workers. To compensate for the hard-working conditions, some over-indulged and, as in any community, there were alcoholics and those who could not handle their liquor. It was in this area that the licensee had an important part to play. In a small settlement, with no opening hours to observe, their ability to control the conduct of their patrons was essential to the settlement's wellbeing. The licensees who filled this difficult role were:

1887-1894	Mrs Violet McArthur
1895-1896	Andrew L Cheyne
1897	William Horne
1898	Andrew L Cheyne
1899	James Hamilton
1900	Mrs W Butler
1901	William Butler
1902	James Hymers

Disaster struck the settlement in October 1896 when the hotel and its contents were destroyed by fire. The fire occurred on the eve of the Bachelors' Ball, the biggest social event of the year, with visitors expected from all around the district. As the Hornes, who ran the hotel, were responsible for the catering, it was reluctantly decided to cancel the ball.

Efforts were made to advise those coming, but the difficulties of getting in touch with most of them meant that many arrived. Some who had sent their ball gowns ahead learned that these, too, had been lost in the fire. To avoid disappointing those who had made the long journey to be present, a dance was organised and, true to the form of the times, dancing went on till daylight. After the dance, William Horne immediately took steps to rebuild, and the new hotel was completed early in 1897.

Everyone who was employed by the company did hard manual labour, often in what would now be completely unacceptable conditions. Almost half of the men worked underground using picks, shovels and explosives in cramped areas and often in wet conditions. Battery-men worked with a constant high level of noise, and everybody was affected by quartz dust, which permeated the entire work site. Silicosis, a debilitating lung disease caused by dust, was a common complaint. It has been said that the two characteristics of quartz mining were the noise of the stamper batteries and the chronic coughs of the workers. Accidents were numerous, and there were some fatalities, including one involving a father and son in 1885.

The rewards for this work varied between seven and eight shillings per day for a surface worker and 10 shillings for underground workers and skilled tradesmen. In comparison, day workers employed on road maintenance by the Lake County Council received seven to eight shillings per day. Continuity of employment was very uncertain. When good stone ran out, the mine closed, often for long periods, leaving employment only for those involved in prospecting. For those not employed, the result was no pay.

Like all small isolated communities, there was a great spirit of friendliness and mutual support. Miles from medical assistance, those with nursing skills helped in times of sickness and childbirth, while those others not so skilled supported in other practical ways. The problems of any one family became the community's problem. When, through sickness or accident, it was necessary to transport someone to hospital, the mine closed and every man helped carry the patient to the road head by stretcher. This task was known to have occupied two days and up to 70 men.

Early in 1883, Fred Evans, the mine manager, having decided to improve the lot of the sick and injured, designed a special wagon to serve as an ambulance. It was three feet wide to suit the narrow pack tracks, could be pulled by a horse and was equipped with brakes. An awning protected the patient, and the bed portion could be detached and carried as a stretcher. The whole wagon was light, yet strong, and could be adjusted to keep the patient level. It was made by the American Carriage Factory of Invercargill at a cost of £45, which was met by a grant of £25 from the hospital committee and the balance from local residents. The ambulance was stored under cover at Edward Fisher's Store on the east bank of the Shotover River, close to the pack bridge at Skippers Point. It was not until 1884 that the ambulance had its first run with a patient, and it went on to prove a great success, speeding up the transfer of patients to hospital, causing much less suffering and no longer requiring the involvement of the whole community.

Entertainment and distractions were few. The billiard room and the hotel were two avenues available to men. Goat shooting and target practice were sports that could be followed, but the usual sports of rugby football and cricket were virtually impossible because there was no suitable area to play on. Bullen Hall was the centre of more passive pastimes. Card evenings, concerts, dances and an occasional ball were all well supported. In a day when almost everyone could sing, recite or play an instrument, organising a concert was fairly easy. The annual Bachelors' Ball was a grand affair, with people coming from as far away as Queenstown and Arrowtown. Dressed in their very best, the participants found plenty of good food and drink, and dancing till daylight. The pleasures of the community were simple but satisfying.

As everything in Bullendale came in by pack horse, the packers formed an important link in the affairs of the community. It was their job to deliver an article, no matter how awkward or heavy. Sheets of galvanised iron, long lengths of timber reaching from over the horses' ears to within a few inches of the ground at their heels were common loads. Iron piping, doors, windows, sashes, chests of tea, bags of sugar and flour, boxes of soap, cases of spirits, wardrobes, kitchen tables, and quartz crushing machinery were all attached to the pack saddles with an expertise acquired by long experience. Packers showed great ingenuity in packing items that appeared impossible to carry by horseback. Those items that were too heavy to be carried in this way were packed on to special sledges. These were pulled by horses yoked in tandem, as the tracks could only accommodate one horse abreast. At the zigzag above Deep Creek and on some parts of the Skippers Creek track, horse power had to be dispensed with and it became a matter of levers, screw jacks, rollers, blocks and tackle. In these circumstances, progress could be measured in inches. Twenty men engaged to move a trolley only 200 yards during a long day's work was not unusual. Winter added a further dimension to a difficult job. Although the pack horses were specially shod to give a grip on icy portions of the track, there were still accidents, with horses, loads or both being lost in steep country. The frustration and financial cost of losing a vital piece of equipment over a cliff after having it specially manufactured and carted the length of the country, or even from overseas, can easily be imagined. People who lived in Bullendale recorded waiting, often in the dark, for the arrival of the pack train with essential supplies and mail. In winter the horses often sported icicles on manes and bellies, evidence of the conditions encountered. It is small wonder that the community regarded packers and their animals with affection.

In 1890 the residents of Bullendale applied to the Southland Education Board for a school to be established in the area. Their application was approved, but was almost certainly subject to the residents contributing towards the teacher's salary. The Board advised that if the number of children fell below a defined minimum, it would contribute at an agreed amount per pupil, with the parents responsible for the shortfall. This situation probably continued over the entire life of the school.

In May 1891 the community held a ball to raise funds for the school. Visitors came from Queenstown, Arrowtown, Frankton and Skippers. The function was entirely successful; entertainment and dancing continued till 6.00 a.m. The school opened in June 1891, with Elizabeth McKersey its first teacher. She was followed by five other teachers before the school closed in May 1902.

The Education Board did not provide a building for the school, and so Bullen Hall was used for the entire period. In May 1897 the Bullendale and Skippers' schools became half-time, with the teacher spending three days per week at Skippers and two at Bullendale. In 1898 committees from both schools agreed to contribute funds to enable the appointment of full-time teachers.

Despite the remoteness and difficulty of access, Bullendale was not neglected by the churches. From the time that full-time clergy were appointed to Queenstown—the Anglican vicar, the Rev. Richard Coffee, in 1869, the Presbyterian Minister, the Rev. Donald Ross, in 1870, and Father John MacKay of the Roman Catholic Church in 1873—regular visits were made to take services at Skippers Point, this being a convenient central point. But as Bullendale developed in numbers, clergy found a need to visit this small settlement high in the mountains. George Bullen reputedly intended to provide a clergyman for the settlement, but this did not eventuate. There were probably not enough local residents to justify an appointment and, besides, there was an acute shortage of clergy throughout New Zealand.

Margaret Glennie, who spent her childhood in Bullendale, recorded the work of two lay members of the Bullendale community who, in the absence of clergy, conducted services and Sunday School. One, an Englishman by the name of Williams, and very likely an Anglican lay reader, conducted Sunday evening services while the other, Furneaux Smith, took Sunday School and delivered a religious tract to every house each Sunday morning. Smith was a Methodist lay preacher, and his dedication to preaching the Gospel earned him the nickname of Holy Smith.

When clergy did arrive to take a service, everybody in the settlement attended, irrespective of their religious persuasion. The Roman Catholic Bishop of Dunedin, Bishop Moran, and the Anglican Bishop of Dunedin, Bishop Nevill, visited Skippers in 1872 and 1874, respectively, and no doubt the faithful from far and wide attended the services on these special occasions. Another clergyman of note to visit Bullendale was the Rev. Alexander Don. A Presbyterian minister for many years, he conducted a mission to the Chinese miners, and Skippers Creek was one of his destinations.

Of all the residents of Bullendale over its 40-years' hey day, it is perhaps Frederick (Fred) Evans who is most inextricably linked with the Bullendale Mine and community, given that he was the manager of the operation for some 30 years. He was born in St. Agnes, Cornwall, in 1831. His father was a mine manager, and it seems likely that Evans had an early introduction to the occupation he followed throughout his life. At the age of 21, he married Catherine Harvey, and there were eight children of the marriage. Catherine died in October 1893, and three of the children predeceased their father. (One son, James H Evans, worked at Bullendale as underground manager.)

In 1864, Evans emigrated to Australia, where he was associated with quartz reefing at the Clunea gold rush. In 1868 he arrived in the Wakatipu, where he took up his position as manager of the Phoenix Mine at Bullendale. Managing a large undertaking like the Phoenix Mine in its remote mountain situation was a complex job. It involved underground mining, with all its associated problems of ventilation, de-watering and ensuring the safety of the miners engaged in this hazardous occupation. When, as often happened in the broken country, the reef was lost, experience was needed to direct the prospecting to find it once more. But whether experience or, in some instances, gut feelings influenced the decisions made, the manager carried the responsibility of endeavouring to maintain payable stone.

The surface workings included crushing, which required heavy equipment. Getting it on site and erected involved almost insuperable difficulties. The loss of plant over bluffs or injury to packers was always possible. Evans was continually writing to the Lake County Council complaining about the state of the tracks, or seeking improvements to make the moving of heavy items easier. Once plant was in place, maintaining it so far from engineering workshops became an added responsibility for the manager.

Evans took much of the initiative and responsibility for the introduction of hydro-electricity to Bullendale, and he deserved credit for pioneering such an important development. He often found the need to make decisions on behalf of the community he led. In June 1885 someone fired the bush near the mine, and Evans organised the fire fighting, which cost the company £50. People management formed an important part of Evans' job. With up to 100 workmen on the payroll at peak times and 100 wives and children, any decisions he made affected many people and an entire community. But for all his problems and responsibilities, it appears that Evans maintained good relations with his workmen and made great efforts to keep the community happy.

In addition to managing the Phoenix Mine, Evans operated the Cornubia Mine in Butchers Creek, a tributary of Skippers Creek, for a short period, but the venture was not a success.

In public life, Evans was very active. He acted as Captain in M Battery of the volunteers for a period and was a trustee of the Wakatipu Hospital. He was a member of the Queenstown Licensing Committee and, from 1868 to 1880, a member of Lake Lodge of Ophir, having previously been a member of Truro Lodge in Cornwall. In May 1887, when the Drill Hall in Beach Street, Queenstown, was opened, Evans sent a generator from Bullendale to light the hall and the adjoining Harp of Erin Hotel for the special occasion. As a civic gesture, carting a generator from Bullendale to Queenstown for one night has few equals.

Evans maintained a house at Bullendale while he was manager and in addition for much of the time farmed 500 acres at Frankton. The property was called Cherry Farm, and his house still stands, though somewhat derelict. It is situated on the Cromwell road, just past the Lake Johnson turnoff.

In 1896, Evans married again, to Elizabeth Knowles. Mrs Knowles was born at Loch Lee, Forfarshire, in 1842. Apparently widowed, she emigrated to New Zealand about 1886, and from 1889 to 1896 was the licensee of the All Nations Hotel at Cardrona. Thereafter, she took up the license of the Reefers Arms Hotel at Maori Point. Her tenure there must have been a brief one before she married Fred Evans. Evans suffered a stroke a couple of years after his retirement from Bullendale and he died on 21 March 1904. Mrs Evans died of cancer on 19 December 1908.

When the mine finally closed down, all those in the settlement had to leave to find work. Leaving was a heartache, as it was impossible to take away many valued possessions because of the difficulty of transporting them. Most families left their homes and most of their furniture, taking with them only their clothes and smaller items of value.

Margaret Glennie's description of the hardship and sadness experienced by her family at this time is telling. Her father, Frank Hughes, was forced to leave his wife and two daughters in Bullendale to take up a job with a contractor in Waipori, but instead of sending money for the fares to Dunedin, he sent word that the contractor had failed. In the middle of winter, Mrs Hughes raffled her sewing machine and kitchen stove to raise enough money for the fares. When she and her daughters reached their destination, they found Frank not only out of work, but with nowhere to live.

Patrick Cotter, who also lived as a child at Bullendale, recorded the fate of Bullen Hall's piano. As it was abandoned, his father arranged to have it carted down to use in his home on the road to The Branches near Skippers Point. It was carted on a sledge, and at each of the river crossings—almost 100 in all—the piano was carried across on a stretcher.

For a long time the abandoned buildings remained intact, but gradually they succumbed to the elements and scavenging, particularly for corrugated iron, which was in short supply as a result of the First World War. Duncan McNichol, who farmed at The Branches, has recalled how he rebuilt the Ballarat Hut on the run, with material recovered from Bullendale: '...the floor timber was off the walls of the manager's house. The other timber and iron came from the hall and other buildings with useable material. The timber was cut to the required lengths, then packed by horse, first day to Skippers Point, next day to The Branches, next day to Shingle Creek and on the fourth day, to the hut site.' After four days packing, these building materials ended up about two miles from where they had been recovered at Bullendale, but on the other side of a feature over 5,000 feet high.

With the passage of time, virtually nothing remains of Bullendale except foundations and heavy items of machinery gradually being covered with the regeneration of the beech forest. The site's isolation and its difficult access mean that few people penetrate this remote spot, and what was once a busy, noisy settlement has returned to its former peace and tranquillity.

Bullendale's Phoenix Hotel, 1887-1902.

Bullendale: workmen at poppet head, 1892.

The hydro-electricity scheme on the left-hand branch of Skippers Creek, Bullendale, was installed in 1886.

Frederick Evans (on right), manager of the Phoenix Mine at Bullendale from 1867-1897, and his wife, Elizabeth, at their home, Cherry Farm, Frankton.

The Road to Skippers

A Summary of Distances and Points of Interest

Page numbers refer to details in history

Kilo–metres		Height above sea level	
		Metres	Feet
0.0	Journey commences at New Zealand Post, Camp Street. Follow Camp and Shotover Streets into Gorge Road.	310	1,016
4.5	Crest of Blow Ho Gully. To the right is a view of Big Beach, site of Sew Hoy's gold dredging success, 1889 (page 73). Cross McChesney's Creek, named after an early publican at Arthurs Point.	419	1,375
5.1	Junction Hotel, Arthurs Point. The hotel is the only survivor of several which operated in this area during the gold rush of the 1860s. The settlement also boasted a number of stores and tradesmen and a police camp (page 64).	396	1,300
5.6	To the left is the start of the walking track to Moonlight, Sefferton and the western bank of the Shotover River. The track is used by a horse-trekking operator.		
5.8	Cross the Shotover River via the Edith Cavell Bridge. The bridge, at a height of 29 metres (94 feet), was named after an English nurse executed by the Germans in Belguim in 1915 for aiding the escape of prisoners-of-war. The area is the scene of the gold discovery by Thomas Arthur and Harry Redfern in November 1862 (page 69), and headquarters for Shotover Jets		

Kilo–metres		Height above sea level	
		Metres	Feet

and rafting operators. Upstream, on the western bank, Oxenbridge Tunnel, an unsucessful mining venture, is now the site of the exciting finale for rafting trips down the river (page 77).

378   1,240

6.4   Pass Racecourse Flat on left—a natural site for horse racing in the days of the rush. On the right is another view of Big Beach, the scene of Sew Hoy's dredging success (page 73). The remains of the Golden Terrace dredge are hidden in the trees on the beach (page 75).

427   1,400

7.5   Packers Arm Restaurant on the right was restored from the remains of the Sportsmens Arms, one of the many hotels on the road to Skippers. The hotel was owned for many years by Patrick Gantley, until his death in 1896. Gantley was an Indian Mutiny veteran and a member of the police force in Queenstown. The hotel's license lapsed in 1899.

7.7   Skippers Road turnoff on the left also leads to Coronet Skifield. The stone buildings on the left were previously store and stables for Julien Bourdeau (page 61). Bourdeau operated a store at Skippers for many years and packed in stores and mail to the area for over 50 years. Until the Skippers Road was completed in 1890, all goods were brought to this point, then broken down intoloads for pack horses. Littles Road on the right leads to Speargrass Flat and Thurlby Domain. The road ahead leads to Arrowtown.

448   1,470

11.7   Beech trees mark the junction of Dan O'Connells Track, a shortcut to Arrowtown completed in 1893 and now no longer used.

762   2,500

Lighthouse Rock, Skippers Road, with the 1863 pack track in the background.
Bob O'Connor

Mt Aurum (7,315′) as seen from Bullendale. 'The Golden Mountain' is said by some to contain the 'mother lode'.
Bob O'Connor

Kilo–metres		Height above sea level	
		Metres	Feet

12.9   On the right, the sealed road leads to Coronet Skifield, a distance of 3.2 kilometres (2 miles).

13.0   On the left, Observation Point and Plane Table.   853   2,800

13.5   Skippers Saddle. A narrow winding road leads down Long Gully and on to Skippers Point, a distance of 16 kilometres (10 miles). The road is recommended for experienced drivers only. Note the 1863 pack track on the right-hand side of the gully.   975   3,200

14.9   Pass through Hells Gates and Heavens Gates, two narrow rock cuttings.

15.4   Pass Lighthouse Rock.   772   2,550

17.8   Across the stream to the right is Balderson's Cottage. After many years working a claim in Deep Creek, Balderson and his wife retired to this cottage, which he described as 'nearer to civilization'.

18.5   Remains of the Welcome Home or Long Gully Hotel (page 63). The first hotel in the area was built by John McArthur in 1863 on the old track on the far side of the gully. It was bought by Peter Bell in 1881 and destroyed by fire in 1889. Bell rebuilt on the present site. The hotel was sold to Henry Lewis, known as Charlie, in 1908 and remained in the family until 1945 when the license lapsed. The building was dismantled and used to build a house at Lake Hayes.   518   1,700

19.2   After a steep descent from the Long Gully Hotel, cross Bells Bridge.

Kilo–metres		Height above sea level	
		Metres	Feet

19.8   Climb Bells Hill to a point where there is a
fine view of the river up and down stream far
below.

20.9   Traverse Pinchers Bluff where, in 1885, the road
was blasted out of a sheer rock face over 200
metres (656 feet) high. Round Devils Elbow and
descend to Deep Creek Bridge.   506   1,660

22.4   Deep Creek Bridge is the departure point
for rafting trips downstream to the Oxenbridge
Tunnel and the Edith Cavell Bridge. It is also the
terminal for jet boating in the upper reaches of
the river.   439   1,440

23.6   Reach Sainsbury Terrace, where there are several
buildings, one a permanent home. There is also
a rest area for tourist buses. Here, visitors can
buy refreshments and look at the mining
museum. A feature of the area is an extremely
deep tail race that discharges into the Shotover
by a tunnel.   527   1,730

24.6   Pass Dredge Slip, where the remains of several
roads eroded by slumping can be seen. A bucket
dredge (page 75) working the river in the first
few years of the century cut away the toe of the
slope, causing slumping.   546   1,790

25.4   To Maori Point Flat, the site of Charlestown
township (page 57). Charlestown was a thriving
town, with a population of over 1,000 at its peak.
It had up to 10 hotels, stores, butchers, bakers, a
police camp and a Magistrate's Court. Being the
only flat area on the Skippers Track, it was the
scene of horse racing and athletic sports.   555   1,820

25.9   A road can be followed down to river level,
where fluming remains in the river from an
attempt by Skippers Ltd to divert the river
from 1930-38 (page 78). On the far bank lies
the remains of the suction dredge which
operated unsuccessfully in 1926/27.

26.7   The road winds through Hakaria Creek, named
after a very early pioneer who struck it rich
(page 69), and on to Maori Point saddle, where
there is a splendid view of river terraces on
the western bank. Many extremely rich claims
were worked here, and the site was also
the scene of the legal battle between Eager and
Grace (page 71). Pass Wong Gongs Terrace,
named after a Chinaman who operated a store
and market garden on this terrace.   655   2,150

28.2   Pass the Blue Slip, a problem area since the road
was built. The road on the right leads to the
Upper Shotover and the Branches Station, a
distance of 16.1 kilometres (10 miles),
(page 80). The road is not recommended for
private cars.   521   1,710

29.5   Pass Bridal Veil Falls to arrive at Skippers
Suspension Bridge (page 17). The bridge
was opened in 1901, 10 years after the road
reached this point. It replaced a pack bridge,
built in 1866 downstream at Londonderry
Creek. The approaches to the old bridge
can be clearly seen (page 13). The present
bridge is 96 metres (316 feet) long and 67
metres (220 feet) high. It is used for
commercial bungy jumping—the
ultimate test for those participating.   503   1,650

Kilo– metres	Height above sea level Metres Feet

29.9   After the steep climb to Burkes Terrace, the
road on the left leads to the old Mt Aurum Station
homestead and the stone school and residence
(1879-1927) now restored by the Conservation
Department. The road on the right leads to
Skippers Cemetery (page 60), then across Sawyers
Creek to the remains of the Otago Hotel
(page 60). Opposite is the site of Aspinall's
sluicing claim (page 72).   570   1,870

The track ahead leads up Skippers Creek for 8
kilometres (5 miles) to Bullendale. Four-wheel-
drive vehicles can negotiate only part of the
distance.

The Bank of New Zealand at Maori Point, 1863. The bank's manager, G M Ross is in the doorway, and the police sergeant is on the right.

The Shotover Gold Dredging Company's suction cutter dredge at Maori Point, 1925. Note the diving bell on the right.

Skippers Ltd, Maori Point, 1932, showing the Shotover diverted into fluming.

Julien Bourdeau—mail carrier to Skippers and storekeeper for over 40 years.

4 Settlements on the Shotover

Maori Point

Maori Point grew up virtually overnight and, for a period, was the main settlement in the area. The scene of very rich gold strikes, it was the only place on the Skippers Track where there was an area of open flat land. The town which grew up was named Charlestown, and for a short time its population probably exceeded 1,000 people. The town's size can be gauged from a report in the *Wakatip Mail* of September 1863 that reported two butchers, one baker, a library, five stores and a hotel on the east side of the river, and a police camp, court house, baker, Bank of New Zealand, three stores and several hotels on the west.

Charlestown was alive with excitement and activity. Riches were won daily, but the easily won gold did not last long and many of the miners departed. By December 1864 the population of the area had dropped to a little over 400. As the population diminished, the storekeepers, hotel keepers and other traders closed their premises and left too. In September 1864 the licenses of seven hotels were cancelled, the licensees having left the district. The names of these shanty hotels and their licensees, who no doubt had for a time profited from the open-handedness of lucky miners, were:

Royal	David Wilson
Inglewood	Samuel Nelson
Harp of Erin	Cornelius Driscoll
Star	Marcus Baker
Coast of Africa	Isadore Haine
Packer	Frederick Clark
Camp	Patrick James Moroney.

Only one hotel survived, The Reefers Arms. It continued to serve as a shelter and place of refreshment for travellers on the road to Skippers until the license lapsed in 1900. The early licensees are not known, but during the hotel's last 11 years of operation it had seven:

1887-89	Madeline Lynch
1890-92	Francis Stone
1893-96	D J L O'Connor
1897	Andrew Cheyne
1898	Charles Dunlop
1899	W Horne
1900	Edgar Sainsbury.

There was a store at Maori Point probably until 1895 when T O'Connor's store and post office burnt down. It is unlikely there would have been sufficient business to re-establish. O'Connor also had the misfortune to lose an earlier store at Stony Creek across the Shotover River in 1887 to the same cause. Another adjoining storekeeper was a Chinaman, Wong Gong, who has given his name to a creek where his business was established about one mile above Maori Point. Wong Gong also had a market garden and catered for his countrymen, who followed the European miners in the area.

Postal facilities were an important service for the miners on the Shotover. A post office opened at Maori Point under Henry W Perryman on 19 September 1863. In 1864, Thomas Goodwin, the licensee of the Diggers Rest, became postmaster. He—and his widow Elizabeth after him—filled the position for 18 years. In 1882 the office was apparently shifted to a local store, but was disrupted in 1895 when O'Connor's store was destroyed by fire. The postal service was restored, but the telephone service closed down. In 1896 the office moved again, this time to The Reefers Arms Hotel, where it remained until the hotel license lapsed in 1900. Postal services continued at Maori Point until 1920. The telephone service was re-established from 1934 until 1941, a period when large-scale mining took place in the area.

On the Shotover, as with all gold rush areas, law and order broke down. With unbelievable riches being won daily, there was plenty of scope for the dishonest and the unscrupulous, who were part of every rush, to enrich themselves. Theft and claim-jumping, often with violence, were common, and the authorities soon saw the need for a police presence in the area. Early in 1863, a station was established with a sergeant and a constable. By 1866, as a result of the reduction in population and mining activity, the station had been reduced to one constable, and before the end of the decade, it closed.

Maori Point, with a court house established in 1863, acted as the administrative centre for justice in the area. In the early days of the rush, there was a great demand for the appointment of a resident magistrate to adjudicate on the numerous mining disputes that arose. The miners rightly sought prompt decisions for disputes, which often involved claims of tremendous value. Having to make the slow journey to Queenstown to get a decision on a dispute that might involve scores, or even hundreds of thousands of dollars in present-day values, was not acceptable. Too often, miners resorted to the other option—force of arms. During the early period of the rush, a resident magistrate was situated at Maori Point, but by 1868 monthly court sessions by a visiting Queenstown magistrate were sufficient to cope with the workload.

A resident clerk handled the routine administrative duties associated with gold fields' activity. These duties included the issue of miners' rights, business and water licenses and the collection of fees and fines promulgated by the Warden's and Magistrate's Courts. The clerk also acted as Receiver of Gold Revenue. His office closed towards the end of the decade.

The only record of a school at Maori Point is for the period 1906-08. This was a time when the Maori Point dredge was operating and so meant that the dredge crew probably had, between them, enough children to justify a school, or were prepared to subsidise the Education Board's contribution.

Today, no sign of Charlestown remains. A small plaque marks the site, reminding those who pass by of the days when the deserted flat was a scene of intense activity.

Skippers Point

The settlement at Skippers Point also began with the 1862 rush. There is no contemporary description as we have for Maori Point, but undoubtedly the Skippers Point settlement was similar in size and makeup. Skippers Creek and the ground surrounding Skippers Point were rich in gold, and for a short time, the population probably exceeded 1,000. However, just two years on, by December 1864, the population was probably only a couple of hundred. Old miners recalled six hotels at Skippers Point, but like those at Maori Point, their life was short. The one hotel that survived was the Otago, owned by Samuel Johnston. Johnston also started a bakery at the Point before there were tracks up the river. Flour had to be carried in by men, and a 4 lb loaf of bread cost 10/-. Despite the exorbitant price, miners besieged the bakery awaiting a batch to be baked. As selling liquor was even more lucrative than bread, Johnston no doubt opened his hotel in the earliest days of the rush.

Johnston continued to combine his bakery business with hotel-keeping for some 30 years. The hotel had a billiard room, and for many years contained the post office. The building was the social centre of the settlement. There, miners could find comfort and refreshment. On his death in 1896, his wife, Eliza, carried on the hotel until 1908 when her son-in-law, John Flynn, took up the license until it lapsed in 1919. Thus, the Johnston family owned the hotel for almost 60 years—the entire life of the business. For many years the old hotel was used as a private residence. It was finally bought by Archie McNichol, the owner of Mt Aurum Station, who intended to demolish it and use the material to build a new homestead. Not being able to do this, and fearing the building might be destroyed by fire as a result of its indiscriminate use, he sold it for removal. The buyer carefully dismantled much of the building, using the material to build a home at Tuckers Beach. However, the part of the old hotel that was built of stone remains, a sad, roofless relic, a reminder of days of activity and success slowly being engulfed by larch trees.

Another reminder of the past is the cemetery on the main terrace. Here rest the pioneers of the area, among them Samuel and Eliza John-ston and successful miner, John Aspinall. Like the remains of the old hotel, the headstones bring back memories of days of joy and sadness and of success and failure. But above all, they stand as a testimony to the courage and tenacity of the pioneers of this area.

The Education Act of 1877 made education free and compulsory for all children. The Government provided buildings, equipment and funding for teachers' salaries based on an agreed amount for each child attending. Under these provisions, Skippers School opened in mid-1879, the first teacher being Charles A Anderson. The school-house, consisting of one classroom and the teacher's residence, was built of local stone at a cost of £300, of which the Education Board contributed £150. The Board also contributed £75 for fittings.

Head teacher at Skippers was not a sought after position, and teachers tended to remain for short periods only. Some 23 teachers taught at Skippers, until it was closed in February 1927. Until the early 1890s the attendances ranged from 25 to 35. After that the numbers dropped to 10 or less, and parents had to contribute financially to keep the school open. After closing, the school remained empty until 1941, when the lessee of the Mt Aurum Station decided to convert the building into a shearing shed. The building was used for this purpose until 1977, at which time the lease of Mt Aurum was terminated. The Department of Conservation decided to restore the building, and this work has now been completed to provide another link with the golden past.

A post office was not opened until 6 August 1867. The first post-master was W Davidson. In 1869 the office shifted to the Otago Hotel, where the Johnston family, Samuel, Elizabeth and daughter Rachel, respectively held the office of postmaster for the 50 years, until 1919, when the hotel license lapsed. However, a postal service continued in the vicinity until 30 April 1952. For most of this time, the office operated from the Mt Aurum Station homestead, except for the period 1927 to 1942, when it was located in several homes, including its former premises, the old Otago Hotel, which by this time was being used as a private residence. A telephone office continued to operate until 1965.

For a short period there was a police presence in the area. In 1864, to cope with the level of mining activity and the criminal and civil problems associated with it, a police station, manned by two constables, was established. The station was closed after a year when the reduction in the population made it possible to administer the area from Maori Point.

Several stores, butchers and bakers supplied the miners' needs. One was owned by Julien Bourdeau, who was born in Montreal of French parents in 1829. He arrived at the Shotover in 1863, and was first engaged in storekeeping at Maori Point before taking over the Skippers Point Hotel. He also ran a general store and packing business. Bourdeau left the area when the rush declined, but returned to the Point in the 1870s, where he re-established his business, but it is not known whether this survived the fire that destroyed his premises in October 1891. However, he did continue his packing operation until his death. Bourdeau at first used pack horses, but replaced these with a two-horse wagon when the Skippers Road opened. Bourdeau was a man of immense strength and vitality, and he carried the mail and stores to Skippers Point for over 40 years despite the difficulties and dangers of the road and the extremes of climate. He experienced financial troubles on several occasions, some as the result of staking unsuccessful miners with food and mining supplies. He carried out numerous contracts in forming and maintaining tracks for the Lake County Council and also contracted for private work such as race formation. For a short time he represented the Shotover Riding on the

County Council, and he worked to the day of his death at the age of 87. His name is inextricably linked with Skippers Point.

Another feature of Skippers Point was its hall and library. Built by public subscription, it was the focal point of the settlement after the hotel, being used for many community activities. Concerts and dances were held at regular intervals, and occasional grand balls attracted people from all over the district. The library was well stocked with leatherbound books, written by the popular authors of the day. Finance came from subscription and from government and Lake County Council subsidies.

Upper Shotover

Several smaller settlements also existed on the Shotover. At Packers Point, on The Branches Road immediately across the river from Skippers Point, a combined store, bakery and butcher's business operated for many years. It served the miners of the Upper Shotover, including those from the Sandhills and The Branches. After Bullendale closed, the Cotter Brothers established their business here, and served customers from the Upper Shotover and from the western banks of the river. At this time, too, the Nugget or Gallant Tipperary quartz mine was operating nearby, but on the western bank of the river. It employed a number of men.

A post office serving the Upper Shotover area opened at the home of William and Ann McLeod, about half way between Packers Point and The Branches, in 1903. It operated till about 1930, and the McLeods ran it for the first 10 years of this period. A story is told about Ann McLeod that is probably true. It is said she went to the Upper Shotover as a bride and, until her death at age 77 in 1915, left her home only once, to attend the opening of the Skippers Bridge in 1901. She is buried in the Skippers cemetery.

The rich terrace claims on the western bank of the Shotover immediately below Skippers Point had a large population during the rush, and the first rates records of the Lake County Council in 1877 show that there were two hotels in the area. Their owners were James Costello and Frank Leyden. As the population at this time must have been no more than two or three score, they must have been hotels in name only. But their presence does indicate an earlier settlement of some size.

Long Gully

The hotel at the bottom of Long Gully formed the centre of the activity in this area, and was a welcome stopping place on the slow journey to Skippers. It was also the social centre for miners in the area, particularly those working on the nearby parts of the eastern side of the Shotover.

Established by John McArthur in 1863, the hotel was situated on the first track to Skippers, on the eastern side of Long Gully. When McArthur died in 1874, his widow, Violet, carried on the business. In 1881, she sold the hotel to Peter Bell and took up the license at Bullendale. The hotel was destroyed by fire in 1889, which must have brought more joy than tears to Bell, as the road to Skippers, which by-passed his hotel, was nearing completion. Bell rebuilt alongside the new road, and carried on as hotelier until his death in 1899, when he froze to death in a water race after falling from his horse. His widow, Honora, took over after his death, until she sold to Henry Lewis in 1908. The Lewis family operated the business until 1945 when the license lapsed.

During the Lewis family ownership, the hotel was again destroyed by fire and rebuilt. After the license lapsed, the building was dismantled and the material used to build a house at Lake Hayes. The hotel was originally named the Travellers Rest, but was changed to Welcome Home under the Lewis's. However, to the locals, it was always the Long Gully Pub.

Today, the site of the hotel is a flat area surrounded by mature trees, with two chimneys standing marking the site of the hotel which, for over 50 years, was a haven to all who travelled the road to Skippers.

Arthurs Point

Arthurs Point, the scene of the first gold strike, has had a residential population ever since that time, though more recently it has been as a suburb of Queenstown. A township quickly sprang up following the discovery of gold, as Arthurs Point was the logical base for communication along both sides of the Shotover River.

The township contained the usual stores, butcher, bakers and traders such as blacksmiths. Equally important were the places of entertainment and amusement, such as the American Bowling Saloon and the various hotels. Three of these were the Shotover, the Rose, Thistle and Shamrock and the Junction. The Junction survives to this day, the only commercial activity in the area apart from tourist operations. James McChesney was licensee of the Junction Hotel from an early date. On his death, his wife Margaret continued the business. Between them, the McChesneys covered a period exceeding 50 years.

The Arthurs Point Post Office had rather a chequered career. It opened on 19 September 1863, but closed again in July 1868 because of the fall-off in population after the end of the main rush. In 1874 it re-opened, with Julien Bourdeau as postmaster. Closed again in 1881, its doors were opened once more in 1889 under the McChesneys of the Junction Hotel. They operated the post office until 1922, after which the hotel's licensees continued to act as postmaster. The office closed for the final time 1931, but the telephone office continued to operate from the hotel until 1964.

There was a school at Arthurs Point from 1893 to 1944. The first teacher was a Miss McDonnell. The school attracted only small numbers of pupils, and for part of its life was half time. The school was situated near the Edith Cavell Bridge, but in 1908 when a new school building was constructed on what is known as the Racecourse Terrace, immediately opposite the present camping ground, the old school and the Millers Flat School on Malaghans Road nearer Arrowtown, which had been half time, were closed and the new Arthurs Point School became a full-time operation. Like the other small schools in the area, teachers did not remain long, and in its first 20 years, the school had 13 teachers.

Arthur's Point, strategically situated at the gateway to the tracks up both sides of the Shotover, had a police station established in 1863, manned by two constables. In 1865 the establishment was changed to one mounted constable, and the station was closed in 1866. Much of the police work originated in the Moke Creek/Moonlight area, which had a significant population in the days of the rush. For a time, Moke Creek boasted a post office, at least one store and a hotel.

Pinchers Bluff, 1897. The road to Skippers was blasted out of a sheer rock face in 1886.

The first Welcome Home Hotel at Long Gully. The building was destroyed by fire.

The Prince Arthur Dredge, Arthurs Point, 1899.

Packing sluice pipes, Shotover Valley.

The Sew Hoy Dredge at Big Beach, Arthurs Point, 1889.

The Talisman Dredge, Lower Shotover, 1891.

5 The Wealth of the Shotover

The first Europeans to discover gold on the Shotover were Thomas Arthur and Harry Redfern. The story of these two men and their discovery in November 1862 is a classic rags to riches tale. Dead broke and dressed in rags, they were engaged by Alfred Duncan at Nokomai, south of the lake, to help shear William Rees' flock of sheep at Queenstown Bay. The pay was to be £1 per hundred sheep shorn, and rations. On arriving at Queenstown Bay, Arthur, hearing there were women living on the station, refused to leave the boat until an old pair of trousers were provided to cover his nakedness. The first Sunday after shearing had commenced (13 November), the two, with a day off, walked up the gully to the Shotover beside the present Edith Cavell bridge. Here, with a pannikin and a butcher's knife, they recovered four ounces of gold. Mad with excitement, they hurried back to the station, stating that they were leaving immediately, as an immense fortune was waiting to be picked up. Although they had undertaken to finish the shearing before fossicking for gold, Rees saw it was hopeless to try and hold them. He gave them flour, tea and sugar, and let them go. A short time later, Vincent Pyke, the Otago Goldfields Secretary, visited Arthur's claim, where he found that Arthur and his three partners had won 200 ounces of gold in eight days. He was shown a tin dish full of gold lying under Arthur's bed. In less than two months, Arthur's party won gold worth £4,000—equivalent in today's values to over $150,000. Arthur's party made no secret of the richness of their claim, and soon the Shotover was swarming with men, officially estimated to number as many as 1,000 at the close of 1862. A month later, another official estimate put the number at 3,000. By June, it was 4,116.

Many struck it rich. At Maori Point, two men, Daniel Allison and Hakaria Haeroa, who swam the river to prospect a likely beach, won 300 ounces of gold before nightfall. They became rich men in a day, for their haul was worth well over £1,000. Pyke records in a mining dispute that a portion of a claim, measuring 5 feet by 12 feet, was valued at £3,000, and subsequent results proved the estimate was not exaggerated. He also recorded a party averaging two pounds weight of gold per day from beach workings. One miner, John Wildredge, wrote to a friend in Dunedin in December 1862 asking for financial assistance in getting his brother and brother-in-law to join him. He told his friend that he could walk into the river up to his waist, dig in his shovel and sometimes bring up five or six ounces of gold, and noted that he 'was making his pile fast—£100 per week'.

Some years later, at a function to honour Richmond Beetham, Resident Magistrate and Gold Warden at Queenstown, a story about a dispute he adjudicated was recounted. The dispute was over an area of wash dirt 10 feet by 3 feet that was a mass of gold. Two parties had claimed the area and, with its riches so obvious, were prepared to take the ground by force of arms. The situation was extremely tense, but Beetham's decision, though not strictly judicial, solved the difficult problem. He ordered 10 men from each party to prepare and, on his starting word, to shovel the wash to their own side.

These are but a few of the stories about riches won in the early days of the rush from beach claims. Those lucky enough to discover an area where the river had been depositing gold over the years were able literally to pick it up. With the aid of a simple gold pan, miners could make fortunes in a very short time. Many others did well, winning more modest returns, but sufficient to set themselves up in business or on a farm, something they previously could never have aspired to. The miners called the Shotover 'The Richest River in the World', but the huge numbers of them on the river meant that the easily won gold was soon exhausted.

With the main rush over, the successful miners left to enjoy their new-found wealth while those who had not been so successful left for other promising fields, drawn by the vision of a rich strike that would enable them to 'make their pile'. Within two years of the initial discovery, the population of the Shotover had fallen to a few hundred and consisted mainly of miners who had a claim producing reasonable returns and who were content to settle at least until the gold petered out. The uppermost thought in every miner's mind was that tomorrow or the next day he would strike a bonanza.

Early on in the rush period, the miners discovered that the west bank of the river between Maori Point and Skippers Point was a rich source of gold. This area was an old bed of the river, but the gold lay under huge deposits of gravel up to 200 feet in depth. The claims were first worked by tunnelling, with some spectacular results.

The richness of one of these claims is well authenticated, as it involved a legal dispute which occupied the courts for a long period and provided plenty of gossip for the locals. A man named Eager had a claim at Pleasant Creek Terrace, which he mined by tunnelling. In 1872 he discovered his neighbours, Grace and party, had driven through from their claim and had been extracting gold for at least a year. Eager sued Grace in the Warden's Court for £12,000 damages, and called in experts and miners to attest to the value of equivalent ground. Grace's bank account showed over £8,000 had been won in nine months. The assessors gave judgement for £8,124 and costs, but Grace appealed to the Supreme Court, and the battle was on. At one stage Eager issued execution and a bailiff seized Grace's claim. Grace and a dozen determined men threw the bailiff out and retook the claim by force of arms. This action provoked further litigation, which went as far as the Court of Appeal. The case may have dragged on even longer if good sense had not at last prevailed. Grace paid out £4,000 in full settlement. The legal costs amounted to several thousand pounds, but such was the wealth of both parties that they cheerfully accepted the fees. In the following three years, £30,000 was taken out of Grace's claim, a sum that would now be equivalent to well over $1,000,000.

This episode shows that miners were not the only ones to profit from gold. Some time after the litigation, Grace's legal adviser in Queenstown, Wesley Turton, built a large home that incorporated expensive imported timbers and included the very best in domestic facilities. This fine home was soon named 'Grace's Folly' by the locals, a reference no doubt to the large sum that Turton would have earned from Grace's claim, and without having to handle a shovel.

Tunnelling was an inefficient means of mining, as it was necessary to timber the shafts, and the areas left for support meant that ground which could be worked was lost. The gravel deposits on the terraces between Maori Point and Skippers Point were, in places, over 150 feet deep, and there was only one way to shift such an immense overburden. This was sluicing with water brought in at high pressure. The introduction of this method was gradual. A sufficient water supply was a big problem, and obtaining it often involved bringing water long distances by water race. Some races exceeded five miles in length, and their cost was very high. A race cut for a Melbourne syndicate to work a claim on Londonderry Terrace took two years to construct. Completed in 1890, it had its source in the Left Hand Branch of Skippers Creek and involved two syphons over the main creek along with four tunnels, the longest being 300 feet and designed to carry 20 heads of water. Its construction cost over £12,000 and was mostly done by contract, with Julien Bourdeau the principal contractor.

Such projects required a lot of capital and could not be undertaken by small parties.

Fortunately, because the ground fell steeply to the Shotover, the usual problem associated with tail races—blockages—was not present. The results of the huge earth-moving operations carried out on the west bank of the Shotover below Skippers Point is evident today, though many of the scars are being screened by the fast-growing larch trees.

One claim at Skippers Point was worked for 50 years, and in some ways its story could stand for that of the Shotover's during its golden days. John Aspinall was one of the early arrivals in Queenstown, in 1862. While prospecting up the west bank of the Shotover River, he found a promising prospect at the junction of the Shotover and Skippers Creek. The early promise was soon confirmed by returns so good that he sent for his brother in Australia. They continued to work the claim by tunnelling, and after two and a half years, brother William, having 'made his pile', returned to Lancashire. John decided to settle near his claim. In 1872 he began sluicing, which involved the employment of labour to cut a race, some of it through solid rock. He set up his own smithy to manufacture pipes, thus overcoming the problem and expense of carting them in by pack horse. Good returns continued, and although by 1894 Aspinall was apparently being dogged by a permanent water shortage, he was still obtaining around two ounces of gold per hour when water was available.

Aspinall married Elizabeth Craigie (originally from Scotland) in 1871, and they raised a family of nine children in a house immediately above the claim. A feature of the two-storeyed house was its large kitchen with its double oven range in the middle of the room, a necessary facility in the long winters. Aspinall maintained a splendid garden and an orchard with a heated green house, an expensive luxury in this remote area. It indicated that the claim continued to live up to its early promise. Bringing in Southland coal by pack horse to heat the green house would have been costly. Another luxury in the home was a piano, laboriously carted in by sledge. A further sign of affluence was the sending of some of the family to boarding school in Invercargill. Aspinall died in 1890 but the claim continued to be worked by the family until 1918, when it was taken over by others. It was not finally worked out until 1937.

Although the easily won gold was soon exhausted, some men stayed on, in some instances for the rest of their working lives. Typical of those who did were Thomas Monk and Neil McInnes. Monk was a Dublin-born Irishman and McInnes a Highland Scot. Together, they worked a claim on the western bank of the Upper Shotover, their access being a cage run on pulleys from a wire rope stretched across the river. The claim was a good one, returning 100 to 150 ounces of gold per annum despite a water shortage problem. The two were well known for their impish sense of humour and the way they treated each other with Old World courtesy. When they first met in the morning, it was always 'Good morning Mr Monk', returned by 'Good morning Mr McInnes'. In old age, having worked the claim for 25 years, they sold out to retire to a cottage in Queenstown. Here they continued a life of harmony and mutual respect until it was broken by the death of McInnes in 1906 at the age of 84.

As time passed, others came on the scene, drawn to the area by stories of wealth won in the past. Most schemes were based on obtaining the gold that everyone was convinced lay in the bed of the Shotover River. It was here that dredging came into use, and again there were some spectacular successes but also many more financial disasters. The Enterprise Company was the first organisation to attempt dredging. It placed a spoon dredge on a pontoon 37 feet by 15 feet in the upper reaches of the river in January 1870. A spoon dredge was merely a large ladle set on the end of a strong arm which, when powered by a block and tackle, could scoop gravel and hopefully gold from the bed of a river by blind dipping. Spoon dredges had very limited success wherever they were used, and the one on the Shotover apparently was no exception to the rule.

But a practical solution to recovering gold from river beds was not far distant. It was bucket dredging powered by steam. The first bucket dredge came into operation on the Clutha River in 1881. Others followed without any spectacular results until a Chinese merchant from Dunedin, Choie Sew Hoy, formed the Big Beach Gold Mining Company in 1888 and placed a dredge on the Shotover River near Arthurs Point. After a disappointing start, Sew Hoy's dredge began to show outstanding returns in 1889. With consistent returns averaging £40 per day, and as much as 29 ounces of gold being won in a day, the dredge began to mirror the successes of the miners of the 1860s rush who struck it rich. In August 1889 shares in the company, originally issued at £9 each, were traded at £120. The prospects were so good that the directors decided to enlarge the capital of the company by offering shares to the public and building three more dredges to work the claim. In a matter of 10 years the original capital of the company of £3,000 had grown to almost £90,000. Only £11,000 of this increase was used to build the three new dredges. In this way Sew Hoy and his fellow promoters capitalised on their original investment and success to date, even if the gold ran out. But the gold did not run out, and the dredges continued to produce returns that triggered a frenzy similar to the rush of the early 1860s. The following figures illustrate the success of Sew Hoy's gold dredging operations:

Period	Gold Revenue	Expenses
March 1889 - November 1890	£4,687	£2,796
Year to November 1891	£ 9,509	£6,895
Year to November 1892	£17,170	£8,796
Year to November 1893	£15,587	£9,175
November 1893 - March 1894	£8,683	£2,914

Miles of river in Otago and Southland were pegged off for dredging claims. Companies were promoted and dredges built. A few of them were extremely successful, fuelling the frenzy, which reached a peak in 1904 when over 150 dredges were operating in Otago and Southland. A greedy and gullible public were taken in on a large scale by unscrupulous promoters, and losses were extensive. Through the early part of this boom, Sew Hoy's company continued to prosper, and at its annual general meeting in November 1892, it was reported that all four dredges were operating, that all liabilities were paid off and that all the proceeds from gold recovered, less working expenses, were available for dividends. The company continued to operate successfully until the claim was worked out in 1898.

Another dredge on the Shotover, the Sandhills Dredge, which worked in the upper reaches of the river, was of interest because its motive power was electricity. The Sandhills Gold Mining Company Ltd had been registered in 1889. The dredge had been prefabricated in Dunedin, and as there was no road from Skippers Point to the working site, transporting its components had not been easy. The difficulties of access and the complete absence of trees for firewood dictated the use of hydro-electricity to power the dredge, and it is generally acknowledged that it was a world's first for this type of motive power. The Sandhills Dredge operated from 1891 to 1895, and although the use of electricity proved successful, the venture itself, from a financial point of view, did not. In 1896 the company wound up, but the dredge continued to operate for some time in the area under new owners until they sold it for use at Millers Flat.

Other dredging ventures existed along the Shotover, although the rocky nature of the river made bucket dredging impossible for long stretches. In 1899 the Arthurs Point Gold Dredging Company was formed to build and operate a dredge. Named the Prince Arthur, the dredge cost £9,000 to build. In 1906 it was reported that the dredge had sunk twice, two companies had gone into liquidation, and the dredge was still under water when a new company was formed. Its dismal record was typical of many dredging companies—started with optimism and finishing with despair.

Maori Point was another part of the river which was dredged. The Maori Point Gold Dredging Company built a dredge at Deep Creek. Transporting the boiler, which weighed 4.5 tons, presented some problems on the narrow road. The dredge worked the river from Deep Creek to Maori Point under two companies from 1901 to 1907 with moderate success. Its lasting claim to fame is Dredge Slip on the Skippers Road. Here, the dredge's operations affected the toe of a slope, which has caused slumping of the road over the years.

The Lower Shotover, nearer to its confluence with the Kawarau River, also came in for attention, even though this part of the river was the final recipient of the millions of tons of overburden that came down the river. In 1891 the Talisman Gold Dredging Company placed a dredge on this section of the river but because of the depth of the gravel had limited success. In 1898 the Golden Terrace Dredging Company Ltd placed a dredge on the same part of the river. In the course of two years it recovered 1,275 ounces of gold, a reasonable return but not a very profitable one.

Despite these unsuccessful ventures, in 1926 the Golden Terrace Extended Company was registered in Invercargill to dredge the same area. The company had a capital of £100,000 and its Chairman, Mr J Holloway, was so carried away with the prospects, he undertook to hand over to charity £1,000 if the shares in the company did not realise £2 on the stock exchange within one year. The dredge was powered by electricity generated at a hydro scheme established at Wye Creek. The scheme still operates today under the ownership of the Otago Central Electric Power Board. Holloway lost his guarantee, but whether he remembered to carry out his undertaking is not known.

After dredging the Lower Shotover flats with only moderate success, the dredge was winched with great difficulty through the gorge to Big Beach, the scene of Sew Hoy's success in 1889. Here its working life ended, and the company was liquidated in 1939. The remains of the dredge can still be seen amongst the willows of Big Beach, one more monument to failed gold mining enterprise on the Shotover.

Although Bullendale was the main quartz mining centre, reefs were worked at other areas in the Shotover Basin. The most successful venture was at Nugget Creek on the west bank of the river, some two miles above Skippers Point. Two reefs, the Nugget and the Cornish, were first worked in the late 1870s. In the 1880s the claims were amalgamated under the name 'The Gallant Tipperary', and profitable results were obtained. The mine employed from 10 to 20 men, depending on circumstances. By 1896 all payable stone had been mined. In the space of 11 years, 11,490 tons of stone had been crushed for a return of 4,392 ounces of gold.

The company went into liquidation in 1898, and the mines were bought by a new unlimited company—the Shotover Quartz Mining Company. Shotover Quartz installed a new 10-head stamper and plant for treatment of the crushings by the cyanide process, operating with varying success until the claim was finally closed down in 1910. The battery remains in its position on the banks of the river as a reminder of an operation that continued for over 30 years.

In 1887 a Melbourne company was formed to mine a reef near Maori Point. It was, of course, named the Maori Point Quartz Mining Company. The initial indications were very promising and a 10-head stamper and other necessary plant were installed. But within a year, the early promise had turned sour, and before the end of 1888 the company was wound up and the plant sold.

About the same time a promising reef was discovered two miles up Butchers Creek, a tributary of Skippers Creek. The prospects were good enough for Fred Evans, manager of the Bullendale mine, to buy the claim and bring in a 10-head stamper to test the ground. By 1890 Evans was satisfied that the reef would not prove successful and he abandoned the operation.

In addition to those described, there have been numerous quartz mining ventures, mostly small, and none of which has proved successful. The Leviathan and Crystal reefs in Sawyers Creek near Skippers Point have been worked from time to time as have others up to modern times. In the time-honoured words of gold prospectors, these claims have shown early promise but, in the fullness of time, the promise was not fulfilled.

Gold has a strange fascination and a lure which defies description. Although over 50 years the Shotover River and its tributaries were mined by pan, cradle, sluicing, tunnelling, crushing and dredging, many people

The Sandhills Cut, Upper Shotover.
Jim Shaw

The Devil's Elbow, Skippers Road. Deep Creek beach, the starting point for many rafting trips, is in the background.
Bob O'Connor

Rock Breaker (1885) looks down on the site of the Bullendale machine house.
Peter De La Mare

The Shotover Gold Company's cutter suction dredge at Maori Point.
Jim Shaw

The remains of Skippers Ltd fluming at Maori Point. The suction dredge is on the beach in the background. Jim Shaw

The last assault on the Shotover River—mining with modern plant. The Maori Point suction dredge is on the right. Jim Shaw

were convinced that riches remained to be won. Although the ventures which followed were not financially successful, they have all left a permanent impression on the landscape and are worth recording.

A short distance upstream from the Edith Cavell Bridge on the western bank is a tunnel some 750 feet long and 14 feet wide driven through solid rock. It is a permanent memorial to the hard work, perseverance and bad luck of the Oxenbridge family. Ten men, five of whom were Oxenbridges, decided in 1906 to tunnel through a headland on the river to divert the water and work the dried-up bed of the river. The work was hard and slow, requiring hand-drilling the rock to place explosives.

After almost three years, the men finally tunnelled through to find that their calculations had been faulty. The tunnel came out four feet too high. Disappointed, but not defeated, the party went to work again to divert the river, and this they eventually achieved. However, their efforts were largely wasted. Having spent many thousands of pounds, their return in gold recovered was minimal. The Oxenbridge party left the claim, broke and downhearted, not knowing that their efforts would later bring delight to many and that the tale of their labours would be related to visitors from around the world. This situation came about with the introduction of commercial rafting and the use of the tunnel as the exciting finale for the journey from Deep Creek to Arthurs Point.

Another claim that permanently changed the landscape was in the Sandhills area in the Upper Shotover, where there was a large loop in the river. The miners decided to make a cut at this point so that they could divert the river and work the dried-up bed. Unlike the Oxenbridge Tunnel, the cut would be through glacial deposits, and sluicing was the method used. The work began in 1926 but was not completed until 1931. The cut was 29 chains long, and some one and a half million cubic yards of material were excavated. Here, too, difficulty was experienced in diverting the river through the cut, and this task was not finally achieved until 1935. Gold was won, but the venture was a financial disaster. The impressive cut between the high banks remains to remind us of another dream that failed.

The year 1925 saw an entirely new scheme to win gold from the Shotover, despite the fact that the area to be covered had been worked and reworked by every means of gold recovery. The Shotover Gold Dredging Company was formed in Melbourne to build and operate a suction cutter dredge, powered by electricity, at Maori Point. The electricity was generated by a pelton wheel, driven by water from a race originating at Stony Creek. The revolving head of the cutter was lowered to the bed of the river, and the gravel dislodged by the cutter drawn up by suction to a gold recovery plant on the deck of the dredge. A long cylinder that could be lowered to the bed of the river to form a diving bell was also installed, on the premise that the miners could use it to determine if gold was trapped in the crevices on the river bed.

The dredge was built in Victoria and re-assembled at Maori Point, and it proved to be a complete failure. The cutter and suction head were designed for silt and fine gravel and could not cope with the large rocks that littered the river bed. The diving bell proved ineffective, while a shortage of water affected the operation. After an extremely short life, the dredge was abandoned on the beach where it had been assembled. It still remains, battered and twisted by floods over the years. And there, in the dry atmosphere and the absence of corrosion, it will probably remain for many years, a pathetic monument to a theory which proved impractical.

With this permanent reminder of a recently failed attempt to recover the gold from the river bed, it is amazing that a further assault was made on the river at Maori Point in 1932. A company, Skippers Ltd, was floated with a capital of £75,000 to divert the Shotover River into steel fluming, thus de-watering much of the river and allowing gold recovery. The fluming consisted of four-feet-square pressed steel tank sections which, with lead washers, could be bolted together in any combination. The fluming eventually stood 16 feet wide with walls four feet high.

It was proposed that once the river was confined to the fluming, the bed would be worked by hydraulic elevating. This method required a good water supply at high pressure, and this was brought in from both sides of the river. However, the water supply proved inadequate, and so a diesel-powered generator and hydro-electricity based on a new dam on Skippers Creek were introduced to drive pumps to ensure an adequate supply. The scheme worked. The river was confined to the fluming and the bed was de-watered to enable gold recovery to proceed. But the amount of gold recovered was insufficient to make the project profitable, and the company went into liquidation in 1937.

If there was one plus to the scheme it was the employment that it provided and the services it used at a time when the country was in the depths of a depression. Some of the fluming sections were recovered and used for building purposes around the district, for which they proved ideal. However, much of the fluming still lies in the river, buckled and twisted by the river in flood, a permanent reminder of another scheme which failed. The concrete dam on Skippers Creek is a further reminder.

The Skippers Ltd scheme was the last big gold mining venture on the

Shotover until more recent times. Some 20 years ago, Ryan Brothers, a Christchurch firm, decided to use their excavating plant to test the river in a manner reminiscent of bucket dredging but much more suitable for dealing with the rough, rocky nature of the river bed. They mounted a massive gold recovery plant on a floating pontoon. The plant was fed by large diesel-powered grabs or excavators that could travel up the river. By this means the entire bed of the river was processed, and with the gold recovery plant many times more efficient than those used in the past, any gold left in the river was won.

Later, another company took over the claim, and after placing two appliances on the river and introducing immense hydraulic diesel-powered excavators as feed machines, covered the entire river. The venture, now finished, was successful. The company reported it had recovered gold worth $26,000,000 in 12 years. When, from time to time, their machinery struck virgin pockets, the returns were as spectacular as those of the 1860s rush, with the gold clearly visible in the wash dirt.

It is now said that at last all the gold in the river has been recovered. But if this is true, the Shotover still has other riches to offer. Since 1867, the Shotover Basin has been used for pastoral farming. The high country of the surrounding area, though rocky and without much topsoil, has large areas where native grasses and tussock grow well. These conditions are good for pasturing sheep, especially such breeds as Merino, which thrive in high country. One limitation on this type of farming, though, is the need for an area at a lower altitude where sheep can be wintered free from permanent snow.

Several sheep stations border the Shotover, the biggest being Coronet Peak, whose western boundary is the eastern bank of the river for almost its entire length. The western side of the river is the boundary for two stations, Ben Lomond and Mt Crichton. But there are two stations more closely associated with the river—Mt Aurum and The Branches. Both have their headquarters on the river and their access up the winding road beside it.

Mt Aurum, with an area of 12,600 hectares (31,135 acres), is based on Skippers Point. It contains within its confines Skippers Point, Bullendale and virtually all the most important mining areas and the visible remains of the mining endeavours. As a sheep station, Mt Aurum was never successful. It had little land suitable for wintering stock, and the difficulty of access made operations expensive. In over 100 years it had 16 lessees, with an average occupancy of three years. On four occasions, the lessee walked off, abandoning the run. In 1977 the Government decided to terminate the pastoral lease but to retain 9,100 hectares (22,487 acres) as a reserve for the gold mining relics and as an area of public enjoyment.

The Branches Station is situated some 16 kilometres (10 miles) above Skippers Point on the Upper Shotover and 48 kilometres (30 miles) from Queenstown. It contains 40,500 hectares (100,000 acres) and, although at a high altitude, has large river flats containing fertile pasture. The flats are open to the sun, enabling the wintering over of the sheep. In 125 years, there have been fewer than 10 lessees, indicating that a station of this size and in this locality can be an economically viable unit. The remoteness and difficulty of access have, to a degree, been overcome by new technology. A landing strip enables fixed-wing aircraft to land at the homestead, and helicopters are common visitors. A private hydro-electricity plant and VHF (very high frequency) telephone to Queenstown are more recent innovations which ease the burden of being at the end of an extremely difficult road.

The road to Skippers and the unique scenery of the area have always appealed to tourists. Since the opening of the bridge at the turn of the century, Skippers has been a tourist destination. Julien Bourdeau was one of the first to carry passengers over this spectacular road in a four-wheeled wagon drawn by two horses. After the Second World War, the introduction of vehicles specially designed by tourist operators for the road has seen a steady increase in the number of people making the trip to Skippers. Today, around 50,000 people travel the road each year.

More modern developments have been rafting, jet boating and the ultimate thrill, bungy jumping, off the Skippers Bridge 67 metres (220 feet) above the river. Most of the rafting starts at Deep Creek, following the river down, working its way through a gorge hemmed in with sheer rock sides, some over 200 metres (620 feet) high, and then negotiating some hair-raising rapids before finally passing through the Oxenbridge Tunnel at Arthurs Point.

Deep Creek is the terminal for a jet boat operation in the upper waters of the river, and some rafting is also done in this section. Other people choose to follow the river and enjoy the scenery in the comfort and speed of a helicopter. Many travel the road in private vehicles to enjoy the scenery and history, and others engage in gold fossicking and tramping.

The road which was built to service the gold mining industry is still serving a similar service, but the gold is now tourism, and the tourists, like the miners of yesterday, come from every corner of the world.

Index

Note: Illustrations and the Points of Interest section are not indexed.